23 Days to Baghdad

U.S. Marine Aviation Combat Element in Iraq, 2003

Colonel Patricia D. Saint
U.S. Marine Corps Reserve (Ret)

Occasional Paper

Reprinted by St. John's Press LLC
Alexandria, Va. 22301
ISBN 978-1-944961-21-3
2016
History Division
United States Marine Corps
Quantico, Virginia
2015

Other Publications in the Occasional Papers Series

Vietnam Histories Workshop: Plenary Session. Jack Shulimson, editor. 1983. 31 pp.

Vietnam Revisited: Conversation with William D. Broyles, Jr. Colonel John G. Miller, USMC, editor. 1984. 48 pp.

Bibliography on Khe Sanh USMC Participation. Commander Ray W. Strubbe, CHC, USNR (Ret), compiler. 1985. 54 pp.

Alligators, Buffaloes, and Bushmasters: The History of the Development of the LVT Through World War II. Major Alfred Dunlop Bailey, USMC (Ret). 1986. 272 pp.

Leadership Lessons and Remembrances from Vietnam. Lieutenant General Herman Nickerson Jr., USMC (Ret). 1988. 93 pp.

The Problems of U.S. Marine Corps Prisoners of War in Korea. James Angus MacDonald Jr. 1988. 295 pp.

John Archer Lejeune, 1869–1942, Register of His Personal Papers. Lieutenant Colonel Merrill L. Bartlett, USMC (Ret). 1988. 123 pp.

To Wake Island and Beyond: Reminiscences. Brigadier General Woodrow M. Kessler, USMC (Ret). 1988. 145 pp.

Thomas Holcomb, 1879–1965, Register of His Personal Papers. Gibson B. Smith. 1988. 229 pp.

Curriculum Evolution, Marine Corps Command and Staff College, 1920–1988. Lieutenant Colonel Donald F. Bittner, USMCR. 1988. 112 pp.

Herringbone Cloak-GI Dagger, Marines of the OSS. Major Robert E. Mattingly, USMC. 1989. 315 pp.

The Journals of Marine Second Lieutenant Henry Bulls Watson, 1845–1848. Charles R. Smith, editor. 1990. 420 pp.

When the Russians Blinked: The U.S. Maritime Response to the Cuban Missile Crisis. Major John M. Young, USMCR. 1990. 246 pp.

Marines in the Mexican War. Gabrielle M. Neufeld Santelli. Charles R. Smith, editor. 1991. 63 pp.

The Development of Amphibious Tactics in the U.S. Navy. General Holland M. Smith, USMC (Ret). 1992. 89 pp.

James Guthrie Harbord, 1866–1947, Register of His Personal Papers. Lieutenant Colonel Merrill L. Bartlett, USMC. 1995. 47 pp.

The Impact of Project 100,000 on the Marine Corps. Captain David A. Dawson, USMC. 1995. 247 pp.

Marine Corps Aircraft: 1913–2000. Major John M. Elliot, USMC (Ret). 2002. 126 pp.

Thomas Holcomb and the Advent of the Marine Corps Defense Battalion, 1936–1941. David J. Ulbrich. 2004. 78 pp.

Marine History Operations in Iraq: Operation Iraq Freedom I, A Catalog of Interviews and Recordings. Lieutenant Colonel Nathan S. Lowrey, USMCR. 2005. 254 pp.

With the 1st Marine Division in Iraq, 2003: No Greater Friend, No Worse Enemy. Lieutenant Colonel Michael S. Groen, USMC. 2006. 413 pp.

Marine Advisors: With the Vietnamese Provincial Reconnaissance Units, 1966–1970. Colonel Andrew R. Finlayson, USMC (Ret). 2009. 72 pp.

Marine Advisors with the Vietnamese Marine Corps: Selected Documents Prepared by the U.S. Marine Advisory Unit, Naval Advisory Group. Charles D. Melson and Wanda J. Renfrow, compliers and editors. 2009. 188 pp.

Operation Millpond: U.S. Marines in Thailand, 1961. Colonel George R. Hofmann Jr., USMC (Ret). 2009. 36 pp.

Close Air Support and the Battle for Khe Sanh. Lieutenant Colonel Shawn P. Callahan, USMC. 2009. 179 pp.

Contents

Foreword

The Marine Corps History Division has undertaken the publication for limited distribution of various studies, theses, compilations, bibliographies, monographs, and memoirs, as well as proceedings at selected workshops, seminars, symposia, and similar colloquia, which the division considers to be of significant value for audiences interested in Marine Corps history. These "occasional papers," which are chosen for their intrinsic worth, must reflect structured research, present a contribution to historical knowledge not readily available in published sources, and reflect original content on the part of the author, compiler, or editor. It is the intent of the division that these occasional papers be distributed to selected institutions, such as service schools, official Department of Defense historical agencies, and directly concerned Marine Corps organizations, so the information contained therein will be available for study and exploitation.

This occasional paper adds to the History Division's examination of U.S. Marines in the Global War on Terrorism and depicts a chronology of Marine aviation events and achievements in Operation Iraqi Freedom during 2003. This publication describes the uniqueness of a scalable Marine air-ground task force and clearly demonstrates its benefits when the 3d Marine Aircraft Wing formed an aviation combat element to better support the ground maneuver element during the rapid 23-day march to Baghdad.

In the period leading up to Operation Iraqi Freedom, our nation expressed a Pearl Harbor–like sentiment that stemmed from the deliberate 11 September 2001 attack on American soil, with parallels between Japanese pilots and the more contemporary terrorist hijackers who took control of civilian airliners. Defense strategists had begun drawing a plausible connection between the 9/11 attacks and Saddam Hussein's suspected arsenal of weapons of mass destruction, thus fueling increased multinational support for an offensive against Iraq. Given this mindset along with military direction, Major General James F. Amos began preparing 3d Marine Aircraft Wing in August 2002 for an air campaign based on ground scheme maneuvers that would logistically and operationally challenge the Marine aviation community—once again.

Recalling the aerial attack on 7 December 1941 at Pearl Harbor, where the U.S. Navy's Pacific Fleet was moored, readers are reminded of numerous accounts of Marines performing beyond expectations during the heat of battle. For example, Marine Aircraft Group 21's commanding officer, Colonel Claude A. Larkin, was returning to Ewa airfield from Honolulu in the early morning hours of 7 December when a low-flying Mitsubishi A6M2 fighter aircraft strafed his 1930 Plymouth. Colonel Larkin survived the initial attack, but suffered several penetrating wounds from subsequent aircraft gunfire. Despite his severe injuries, he managed to direct the defense of the smoke-filled airfield to minimize damage to aircraft and personnel. Colonel Larkin ordered his Marines to set up additional security around the airfield and to redistribute stockpiled ammunition to prevent a massive explosion, while conducting rescue and recovery efforts for fallen and wounded Marines. Meanwhile, fighter pilots like Captain Robert E. Galer unsuccessfully tried to reach their cockpits during the attack. They were handed rifles instead and ordered to "fire in volleys," taking refuge inside a huge hole that had been dug for a base swimming pool.

In this book, you will read accounts of Marines of all ranks and from all squadrons performing similar acts during Operation Iraqi Freedom. This holds true, particularly during the Opening Gambit, the Battle of an-Nasiriyah, the rescue of Jessica D. Lynch, and the Task Force Tripoli prisoner-of-war recovery mission. In retrospect, the aviation combat element strategy allowed the air wing to operate as a separate maneuver element with Marines controlling Marine aircraft in a joint arena. Ultimately, Iraqi combatants were forced to make two basic choices—remain in position and fight or flee and become a convenient target of opportunity.

C.P. Neimeyer

Dr. Charles P. Neimeyer
Director of Marine Corps History

Preface

More than a decade after Major General James F. Amos led the 3d Marine Aircraft Wing (3d MAW) in air combat during Operation Iraqi Freedom, this occasional paper joins the History Division's discussion of U.S. Marines in the Global War on Terrorism. More importantly, this work pays tribute to all those who served, sacrificed, and shared their stories.

This historical account captures the monumental administrative, logistical, and operational planning and preparation activities leading up to the launch of the air campaign that began on 19 March 2003, while also providing an event-driven timeline during the 400-mile journey from Kuwait to Iraq's capital. It highlights Marine aviation with its full spectrum of doctrinal capabilities and 3d MAW's 400-plus tactical aircraft, along with the rationale for forming an aviation combat element that would also operate as a separate maneuver element to better support ground forces during their rapid advancement to Baghdad. Although unconventional, this strategy was critical for complementing the need for speed and surprise to support the group troops—thus the advance became the eyes for the 1st Marine Division.

This book portrays repeated examples of small unit leadership, camaraderie, and "can-do" spirit within every squadron and aircraft group that collectively contributed to the 3d MAW's successes during the march to Baghdad. Until the end of the air campaign, aircrews tirelessly flew planned and on-call fixed- and rotary-wing sorties during darkness and through harrowing sandstorms to deliver troops, medical evacuations, supplies, and ordnance on target, while maintenance crews and aviation logisticians retained high operational readiness rates despite austere conditions. Given the solid relationship and operational agreement in place between General Amos and U.S. Air Force Lieutenant General T. Michael Moseley, the Coalition air component commander, Marine aviation was allowed to directly manage Marine assets with joint Air Force accountability. Marine wing support squadrons joined with air command and control Marines to keep pace with tactical air wing operations by leapfrogging primarily in tactical convoys to forward operating bases and forward arming and refueling points. When air combat operations ended, Marine aviation flew 22 percent of Coalition missions and lost only six rotary aircraft to Class A mishaps or crashes. The 15,000 Marines who served under General Amos's command—whether they have retired, left active duty, or still wear the uniform—will forever remember receiving the Presidential Unit Citation on 3 November 2003 and witnessing "Marines taking care of Marines" during Operation Iraqi Freedom.

My initial appreciation of Marine aviation and desire to serve began early in my childhood with my father, Marine First Lieutenant Irvin J. Dillon, telling stories about flying Vought F4-U Corsairs during World War II, and with my mother, Imogene B. Dillon, instilling a deep-rooted sense of patriotism. As a company grade officer, my military aviation career included assignments as a tactical air controller with Marine Air Control Group 28 and as an aviation logistician with Marine Aircraft Group 14; as a reservist, I served as an officer-in-charge of mobilization readiness at Marine Corps Air Station Cherry Point, North Carolina. Collectively, these roles provided a firm foundation for understanding the complexities and addressing the challenges and successes of Marine aviation during Operation Iraqi Freedom. Asked by retired Colonel Richard D. Camp Jr. to write the Marine aviation story for History Division, I was mobilized to active duty in 2006.

This occasional paper was made possible by the unwavering support and encouragement of Dr. Charles P. Neimeyer, director of the Marine Corps History Division, along with superb editorial guidance from Charles D. Melson, chief historian of the Marine Corps History Division. My sincere thanks extends to other History Division staff for providing peer and editorial reviews: retired Colonel Stephen S. Evans; Colonel Nathan S. Lowrey; retired Lieutenant Colonel David A. Benhoff; Major C. Cameron Wilson; Annette D. Amerman; Wanda J. Renfrow; James M. Caiella; Kenneth H. Williams; Shawn H. Vreeland; and Angela J. Anderson. Fellow Marine Colonels Teresa L. Dillon, Jincy L. Hayes, and Ronnell R. McFarland also provided reviews, while aviation-focused field historians Dr. Fred H. Allison, Colonel Jeffrey Acosta, and Colonel Michael D. Visconage shared numerous oral interviews and photographs. My genuine appreciation also extends to the graphic design talents of W. Stephen Hill, who designed and laid out the book, and Vincent J. Martinez, who helped transform a collection of photos into a visual storyboard.

Finally, I want to thank my family, retired U.S. Navy Commander William K. Saint and my sons, Thomas J. and Patrick D. Saint, as well as my sister, Letitia A. Moscrip, for truly understanding and supporting my call to duty.

Undoubtedly, it was a genuine honor to write this occasional paper, particularly collecting, re-interviewing, and sharing the stories of those Marines and sailors who were part of the 3d MAW. I am grateful for the publication of this book, thus adding another chapter to the proud legacy of Marine aviation.

Patricia D. Saint
Colonel, U.S. Marine Corps Reserve (Retired)

Prologue

Defending America's Homeland

Freedom, itself, was attacked this morning by a faceless coward, and freedom will be defended.

—President George W. Bush, 11 September 2001[1]

First Responders

Nearly 60 years after Japanese bomber and fighter pilots attacked the U.S. naval base at Pearl Harbor, 19 Islamist terrorists hijacked four civilian aircraft shortly after takeoff, murdering flight crews and replacing licensed aviators with suicide pilots. The hijackers seized control of the cockpits, deviated from flight plans, and redeployed the passenger planes as deadly missiles—with innocent civilians on board—toward predetermined targets, specifically the twin towers of the World Trade Center in New York City and the Pentagon near the District of Columbia.* As our nation witnessed the series of surprise aerial attacks unfold on 11 September 2001, the U.S. Marine Corps and its sister services would soon be called on to defend America once again.**

Inside the Pentagon, where many military personnel had just begun reporting for work, American Airlines Flight 77 crashed into the east side of the building, engulfing it in flames. Not surprisingly, Marines reacted immediately alongside other military personnel, assuming leadership roles, assisting, directing, and rescuing others from the burning building. Unaware at first that a plane had crashed into the Pentagon, Colonel Susan G. Sweatt, a fiscal officer and former tactical air controller, made lifesaving decisions as she quickly led her civilian staff to safety away from a smoke-filled office suite to the outside grounds where they joined a dazed workforce of 18,000 military and civilian personnel. In the midst of a mass evacuation that overflowed into nearby Arlington, Virginia, neighborhoods, Colonel Sweatt recalled the depth of kindness local residents displayed by offering food, water, telephone access, and hospitality for displaced and stranded victims and extending repeated gestures of human compassion.[2]

U.S. National Archives identifier 6604922

A view of the wreckage at the World Trade Center site on 15 September 2001, just four days after the terrorist attacks of 9/11. An American flag can be seen draped from the skyscraper known as Two World Financial Center, which was damaged by falling debris from the collapse of the twin towers.

Lance Corporals Dustin P. Schuetz and Michael Vera were thrown to the floor by the blast, lying beneath dozens of manuals, books, and debris. Uninjured, they evacuated their offices and joined other rescuers on a lower floor where the scene was more gruesome; there, they removed more seriously injured personnel trapped in mounds of rubble. Relying on natural instincts and survival skills, rescuers rolled around in pools of standing

*By midmorning, three aircraft had purposely crashed into their intended targets, demolishing both World Trade Center towers and the east side of the Pentagon near the fourth corridor. A fourth aircraft crashed in rural Pennsylvania while en route to its intended target—the U.S. Capitol building—which was spared from destruction due to heroic passengers who struggled with their hijackers on board United Airlines Flight 93.

**After the 11 September 2001 terrorist attacks, the day has been referred to as September 11th or 9/11, which signifies a universal fast dial telephone code (911) used to summon emergency police, fire, or medical assistance.

Photo by TSgt Cedric H. Rudisill, USAF. Defense Imagery 010914-F-8006R-006

An aerial view of the area where American Airlines Flight 77 slammed into the Pentagon on 11 September 2001. Ironically, this section was in the final stages of a renovation to reinforce and update the building. The huge American flag visible to the right of the damaged area is a garrison flag sent by the U.S. Army Band at nearby Fort Myer, Virginia.

water to refresh their bodies from the intense heat of the raging fire. Unconcerned for their own safety, they continued searching for more trapped victims, estimating that they may have saved a dozen lives before firefighters arrived on site.[3]

On another side of the building, Major Bradley H. Shumaker was drinking a cup of coffee and watching a television monitor near the Pentagon Joint Military Command Center* when he first heard the breaking news about the World Trade Center collapse. Serving as a senior crisis action team member on the joint operations staff, Major Shumaker overheard an incoming phone call from the Federal Aviation Agency (FAA) that reported another aircraft hijacking in progress. At the same time, he recalled an annoying grumbling noise, similar to the roar of a faulty industrial air conditioner, followed by piercing alarms moments later. Quickly deducing that another attack had occurred, the major assembled a group to field executive-level communications. Eventually, the team left the smoldering building and established a more permanent alternate command post in a different location, flying nap-of-the-earth by helicopter up the Potomac River with no other aircraft in sight.[4]

In a terrifying span of 1 hour and 20 minutes, all four aircraft had crashed in New York, Virginia, and Pennsylvania, killing 256 passengers and crew members upon impact, while the aftermath from the building fires and collapses subsequently claimed an additional estimated 3,000 people, surpassing the number of lives lost during the Pearl Harbor attack.[5] Ironically, the 9/11 surprise attack occurred on the 60th anniversary of the Pentagon's groundbreaking ceremony.

U.S. Airspace Temporarily Shuts Down

Meanwhile, as smoke billowed from the Pentagon, other government agencies were scrambling in response to the series of attacks. Former Chief Warrant Officer Mark D. St. Amand reported for his morning shift as a civilian air traffic controller at the Washington air traffic control center in Leesburg, Virginia, unaware that he would face one of the most challenging days of his professional career. At 0800, he assumed temporary supervisory duties as the controller-in-charge, while his boss took a short coffee break. The FAA had already forwarded its first warning of a terrorist attack at 0838 to the Northeast Air Defense Sector within the North American Aerospace Defense Command when the transponder on American Airlines Flight 11, the first aircraft to crash, no longer provided a signal.[6] At that time, St. Amand was controlling seven different sectors that equated to about

*The Pentagon's command center operated as a 24-hour crisis action site to inform the president of the United States and the chairman of the Joint Chiefs of Staff of current operations.

30–45 aircraft; although the information certainly piqued his interest, he implemented no special procedures other than to advise other controllers of the situation. News of the second crash, however, generated a heightened tempo; a situation of this nature had never been rehearsed during mandated security control of air traffic and air navigational aid drills. Further, in an unprecedented decision in the history of civil aviation, the FAA completely shut down U.S. airspace to commercial and private aircraft—except for military flights.*

St. Amand's controllers rapidly became saturated with coordination decisions related to diverting, landing, and parking airplanes at any airport with space available. As he remembered:

> We would arrange for aircraft to divert to Norfolk or Raleigh, and then would receive notification that those airports had reached their traffic capacity. Some aircraft had to dump fuel because they were too heavy and exceeded landing weights at specific airports. We continuously coordinated with adjacent centers about approach control and whether to park aircraft on runways or at nearby hangars. About two to three hours later, all aircraft in our sector had been successfully parked, yet it was eerie to see blank radar scopes—just empty screens and no airplanes flying in the sky. The airspace was empty.[7]

From an airline pilot's perspective, US Airways Captain William K. Saint Jr., a retired U.S. Navy commander, was holding in the number one position to take off from runway 28-Right at Pittsburgh International Airport. With no knowledge of any terrorist attacks, the McDonnell Douglas MD-80 captain finished his departure checklist and waited for final clearance from air traffic control at the same time the second hijacking was underway. Noting that it was taking an unusually long time for a response from the control tower, Saint reacted immediately when a controller calmly yet decisively stated a most unusual command, "The airport is closed, return to the gate." Air traffic controllers relayed this message across the country, and just like many other pilots who heard the same command that morning, Saint, a former Lockheed P-3 Orion pilot, did not question the order. Instead, he promptly returned the aircraft to the jetway and helped passengers deplane, despite being unclear as to the reason for sudden and massive grounding of flights. Regardless, once inside the terminal, people were quickly exiting the airport complex in reaction to reports about a hijacked airplane that was headed for the Pittsburgh airport. He recalled the scene:

> Without any ground transportation, I walked a couple of miles to a nearby hotel where several other airline crew members and stranded passengers stood in the lobby, searching for vacant rooms. Later that afternoon, I was bothered by the stillness with no familiar roar of jet engines or the unique smell of aviation fuel. The only activity at the airport was a couple of Pittsburgh Air National Guard [Boeing] KC-135 [Stratotanker] aircraft that I later learned were flown to refuel fighter aircraft assigned to patrol the skies over Washington, DC.[8]

By sunset, the FAA and its many air controllers like St. Amand had safely grounded more than 4,500 airplanes. After three days on standby status, U.S. airspace reopened. Saint was among the first pilots who began flying a limited schedule, but this was initially without any commercial flights into or out of the Washington, DC, area.[9]

Hornets on Patrol

Within 24 hours of the attacks, Marine aviation flew one of its first missions defending America's homeland when Marine Fighter Attack Squadron 321 (VMFA-321), a reserve McDonnell Douglas F/A-18 Hornet squadron, was called upon to fly combat air patrols over the nation's capital in support of Operation Noble Eagle.* Based at Andrews Air Force Base in eastern Maryland, the commanding officer, Lieutenant Colonel Robert A. Ballard, placed the squadron on standby status and ral-

*A civilian air controller, Benedict L. Sliney, ultimately made the critical decision from Herndon, Virginia. Ironically, Sliney, a former New York lawyer who had once sued the FAA on behalf of air controllers, was just beginning his first day on the job as national operations manager—the "big boss" of the air traffic system—when reports of the hijackings began flooding his terminal.

*Operation Noble Eagle involved many services, and by Friday, 14 September, Capt Ralph E. Bally, USN, arrived in New York City on board the USS *Comfort* (T-AH-20), a U.S. Navy hospital ship to support search, rescue, and recovery efforts. As a member of a mental health team, Capt Bally distributed pamphlets that identified symptoms of anxiety and posttraumatic stress disorder; offered a formal debriefing for police officers, firefighters, emergency medical technicians, and other rescue workers; and provided humanitarian relief until 1 October.

lied his Marines to report to the hangar.* Fully anticipating a mission, proactive maintenance crews had already begun refueling fighter-attack aircraft, loading ordnance payloads with 20mm cannon rounds and a few AIM-9 Sidewinder air-to-air missiles borrowed from the District of Columbia Air National Guard's 121st Fighter Squadron. Coordinating with the Northeast Air Defense Sector, Colonel Ballard briefed his aircrews on the distinct ad hoc rules of engagement, authorizing pilots to identify and target any aircraft that penetrated the 25-mile perimeter around the nation's capital. Because several of the pilots were also commercial airline pilots, this newfound direction punctuated the seriousness of the overall situation.** By the end of the day, approximately 70 VMFA-321 personnel had reported to the squadron, including 15 pilots from various regions of the country who camped out in the squadron hangar, sleeping on the floor or in ready room briefing chairs. They flew over the U.S. Capitol until relieved by an Air Force squadron the following day.[10]

Retaliation: Operation Enduring Freedom

On that 9/11 morning, President George W. Bush was touring an elementary school classroom in Sarasota, Florida, when he was informed of the first airplane crash. After a brief moment of silence, Secret Service agents whisked him away to a nearby airport, and by 0955 the president was airborne on Air Force One en route to an alternate commander-in-chief command post.[11] Later that afternoon, the National Security Council conveyed its belief that Osama bin Laden and his network of al-Qaeda terrorists were responsible for the attacks.[12] Although his words were brief, President Bush addressed the American people in a two-minute speech from Barksdale Air Force Base in Louisiana, reassuring them of his commitment to "hunt down and punish those responsible for these cowardly acts."[13]

At the White House the next day, he began making the legal, economic, diplomatic, and military decisions necessary to act on that commitment, beginning with passage of Senate Joint Resolution 23. Four days later, it became Public Law 107-40 and authorized the use of U.S. Armed Forces against those responsible for the recent attacks launched against the United States.[14] Central Intelligence Agency briefings further profiled the 9/11 suicide pilots as Islamist terrorists from the al-Qaeda network, safeguarded by Afghanistan's extremist Taliban regime.[15] As more information surfaced, it became evident that the hijackers not only intended to instill fear in the American homeland, but they had a long-term commitment to disrupting the economic, military, and political pillars of the United States—in essence, the American way of life.

With the president's legal and economic actions in place, orchestrating a military response against the terrorist network proved difficult because the hijackers and those associated with the 9/11 attacks wore no military uniforms to signify a specific country of origin. Additionally, a polished operational plan for an Afghanistan campaign was nonexistent.* Regardless, Secretary of Defense Donald H. Rumsfeld directed the U.S. Central Command (CentCom) commander, U.S. Army General Tommy R. Franks, to prepare "credible military options." Franks's staff began formulating a joint service contingency plan to neutralize the terrorist threat posed by the Taliban regime and al-Qaeda radicals in Afghanistan.[16] He reviewed an existing contingency operational plan written three years earlier by his predecessor, Marine General Anthony C. Zinni.** This plan focused on Iraq, however, rather than Afghanistan.[17] General Zinni's Operational Plan 1003-98 outlined an Iraq invasion with a gradual Middle East presence over a 10-year period, requiring a force of 380,000 troops with logistical support for a potential operation in Afghanistan.

CentCom planners at MacDill Air Force Base

*The squadron earned its nickname—"Hell's Angels"—while flying F-4U Corsairs on combat missions in the South Pacific during World War II. In September 2004, the squadron was decommissioned after 61 years of service, and it proudly added the events and unit contributions surrounding 9/11 to its final command chronology.

**The rules of engagement were direct, yet initially unsettling, especially for several of the squadron pilots who were civilian airline pilots. During this mission, they may have been ordered to shoot down an airliner similar to one they flew commercially.

*Just two weeks after the attack, President Bush signed Executive Order 13224, which authorized the United States to identify individuals, businesses, charities, and extremist groups engaged in terrorist activities and to curtail future terrorist funding by freezing bank account access for Afghanistan's Taliban regime if linked to the attacks.

**Gen Zinni began his 35-year career as an infantry battalion advisor to the Vietnamese Marine Corps, and he retired in September 2000 as the commander in chief of CentCom.

in Tampa, Florida, worked arduously for days to consolidate all options and to formulate a broad, phased campaign in Afghanistan in order to accomplish the following goals:

1. set conditions and build forces to provide the national command authority with credible military options;
2. conduct initial combat missions for follow-on operations;
3. conduct decisive combat operations; and
4. build a coalition network and provide humanitarian and civil affairs assistance.[18]

On 20 September, President Bush addressed a joint session of Congress and the American people in a 41-minute speech in which he identified the 9/11 terrorists as members of the al-Qaeda organization—the same group responsible for bombing American embassies in Tanzania and Kenya and for attacking the destroyer USS *Cole* (DDG 67) in the harbor of Aden, Yemen, a year earlier. The president outlined several measures to minimize the threat of another attack and announced the creation of the Department of Homeland Defense.[19] Seventeen days later, Operation Enduring Freedom (OEF) became the forward deployment in a global war against terrorism.

On 7 October 2001, the United States launched an intensive, three-month retaliation operation in Afghanistan against the Taliban, targeting the al-Qaeda network and its mastermind, Osama bin Laden.* The opening offensive operations included a mix of air strikes from land-based bomber aircraft and carrier-based fighters along with strikes by General Dynamics BGM-109 Tomahawk cruise missiles launched from U.S. and British ships and submarines. In the early morning hours of 18 October, VMFA-251 joined the OEF effort.[20] Based at Marine Corps Air Station Beaufort, South Carolina, the F/A-18 Hornet squadron flew its combat mission alongside two U.S. Navy Grumman F-14 Tomcat fighters from the USS *Theodore Roosevelt* (CVN 71), which was the flagship of one of the four Navy battle groups in the North Arabian Sea. Just before midnight, squadron personnel lined the flight deck where the commanding officer of VMFA-251, Lieutenant Colonel Raymond C. Damm, and his wingman, Captain Simon M. Doran, led an initial air strike mission. Colonel Damm inspected his aircraft configured with two 2,000-pound joint direct attack munition precision-guided bombs, then carefully checked the squadron colors behind his seat where Sergeant Major Sergio J. Estrada Jr. had placed them a few hours earlier, both knowing that the return of the colors to battle would instill unit pride for the sailors and approximately 195 Marines on board the carrier. During the eight-hour mission, Damm and Doran bombed targets almost 500 miles farther inland than briefed and destroyed a key bridge in northern Afghanistan used to resupply the Taliban or al-Qaeda near Mazãr-e Sharif—without a doubt the mission was successful.[21]

The terrorist attacks of 9/11 cast that day in history, fueling anger and creating a Pearl Harbor–like sentiment among Americans, which, in turn, initially generated military, government, and public support for OEF in Afghanistan. By early 2002, the operation included a coalition of allies that included the United Kingdom, Spain, Australia, Poland, Denmark, and Italy with support eventually growing to more than 68 nations; 27 countries maintained representatives at CentCom's forward headquarters in Bahrain.[22] Joint Coalition efforts temporarily thwarted Taliban and al-Qaeda aggression in Afghanistan. Although retaliatory air strikes and joint special operations helped overthrow the country's repressive Taliban regime and severely disrupted the al-Qaeda network, Coalition military efforts were unsuccessful in capturing the terrorist ringleader bin Laden.* Just as in Afghanistan, Marine aviation would play a significant yet somewhat different role supporting ground forces during Operation Iraqi Freedom in March 2003.

---

*The son of a wealthy Yemeni businessman, Osama bin Laden was the leader of the international terrorist organization al-Qaeda. In 1998, bin Laden issued a *fatwa* (religious ruling) under the banner of the International Islamic Front for Jihad against Jews and Crusaders stating it was the duty of all Muslims to kill Americans and their allies.

*For the most part, the Bush administration believed the al-Qaeda network, particularly Osama bin Laden, was responsible for orchestrating the 11 September 2001 terrorist attacks. The Taliban regime safeguarded al-Qaeda terrorists, and it was believed bin Laden was hiding in Afghanistan soon after the attacks.

Part I

PREPARATION

Chapter 1

Building Relationships for War

Expanding the Global War on Terrorism

The weekend before American Airlines Flight 77 crashed into the Pentagon on 11 September 2001, Major General James F. Amos, Assistant Deputy Commandant for Aviation, had begun relocating most of the aviation division from an office suite directly above the building's west side. It was a scheduled move to the opposite side of the sprawling complex, and fortunately none of the Marines or civilians in the organization were seriously injured or killed during the resulting explosion that rocked the Pentagon.[1] Eighteen months later, General Amos was no longer advising the Commandant of the Marine Corps on all aviation matters or making strategic decisions from his Pentagon office. Instead, the general was directing one of the largest Marine aircraft wing deployments since Operations Desert Shield and Desert Storm in 1990–91. This time, Iraqi dictator Saddam Hussein was believed by American leaders to be actively producing weapons of mass destruction and harboring al-Qaeda terrorists.*

On 19 March 2003, a U.S.-led Coalition spearheaded a major shift in military focus from Operation Enduring Freedom (OEF) in Afghanistan to a much larger military offensive—Operation Iraqi Freedom (OIF). As an integral part of this multinational effort, General Amos commanded the reinforced 3d Marine Aircraft Wing (3d MAW) and led more than 15,000 Marines into combat, while overseeing 435 tactical aircraft staged at air bases in Kuwait, Saudi Arabia, and Bahrain and those on board naval ships.[2] Following Marine air-ground task force (MAGTF) doctrine, General Amos organized the air wing to support the ground scheme of maneuver, yet he also molded an air campaign that employed the aviation combat element in an unusual role as a separate aviation maneuver element. Operating in dual roles, 3d MAW conducted armed reconnaissance, assault support, and close air support missions as Marines moved toward Baghdad while simultaneously engaged in deep air strike missions, seeking and destroying enemy artillery, command-and-control sites, and tank columns well ahead of ground forces. In essence, the latter role enabled the wing to become the eyes of the 1st Marine Division (1st MarDiv), providing ground commanders with real-time snapshots of enemy activity on the battlefield and, in turn, greatly influencing tactical decision making.

Photo by Sgt Nicholas S. Hizer. Defense Imagery 020916-M-UE267-001

Directing one of the largest unit and aircraft deployments since the Gulf War, MajGen James F. Amos, commanding general of the 3d MAW, deployed Marine aviation in a supporting role and also as a wing maneuver element during Operation Iraqi Freedom.

*Concern over Iraqi weapons of mass destruction dated back to Saddam's use of chemical weapons against the Iranians and the Kurds in the 1980s. This concern was formalized under President William J. Clinton's administration, when he signed House Resolution 4655 into law on 31 October 1998. Referred to as the Iraqi Liberation Act, the resolution outlined support for a brighter future for the oppressed people of Iraq.

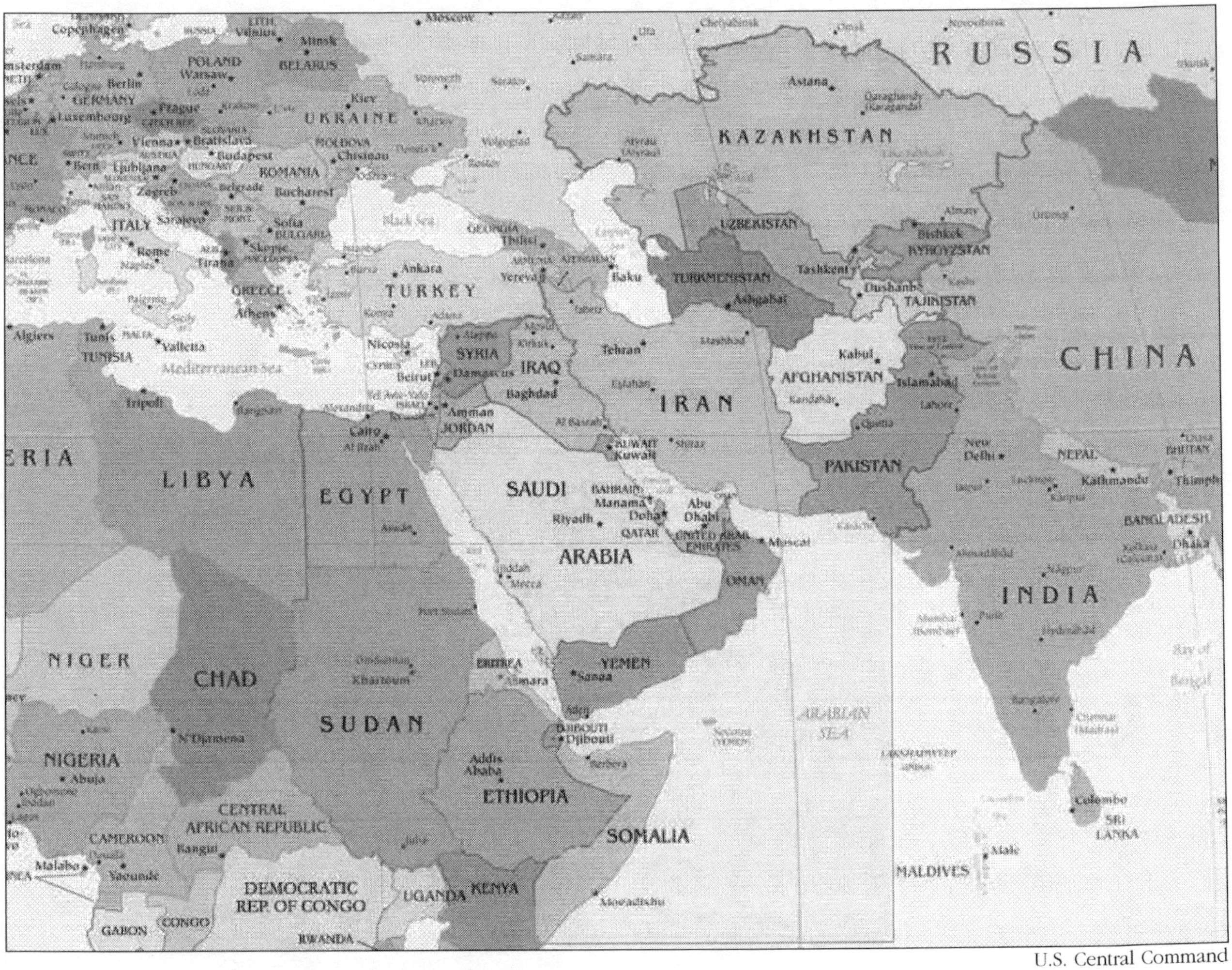

U.S. Central Command

U.S. Central Command's area of responsibility in the Middle East.

The full range of Marine Corps aviation capabilities emerged with the wing maneuver element.[3] Despite never-ending operational and logistical challenges, General Amos and the wing extended air operations farther north than planned and participated in the historic fall of Baghdad. It is this collective account—shared by wing commanders, pilots and aircrew, air controllers, aircraft maintainers, logisticians, and aviation ground support staffs—that highlights the achievements, sacrifices, and spirit of Marine aviation, thus adding another chapter to its story.

Linking al-Qaeda to Iraq

Understanding the rationale behind the transition from a military effort rooting out terrorists in Afghanistan to a stake deep inside Iraq is equally as important as gaining insight about OIF itself. At a time when recollection of the 9/11 terrorist attacks had begun to fade, expanding the war on terrorism to Iraq was not a surprise for many military and civilian analysts who had spent years monitoring terrorist activities, specifically those studying al-Qaeda and the Taliban. Based on long-standing rumors of Saddam Hussein's nuclear program, some analysts expressed grave concerns about Iraq's alleged stockpile of weapons of mass destruction that, if obtained by terrorist groups or rogue nations, could result in another terrorist attack on American soil.[4]

In retrospect, endorsement of an invasion of Iraq from the international community appeared mixed. Israel, Kuwait, and Qatar voiced strong regional commitment with backing from the United Kingdom, Australia, Spain, and Poland; however, Germany and France opposed hostilities against Saddam's regime, publically expressing their negative sentiments.[5] Even though the U.S. government solicited approval from the United Nations (UN) to use military force to carry out UN Security Coun-

Origin of U.S. Central Command

More than 60 years ago, the U.S. military relied on the National Security Act of 1947 to divide the world map into geographical regions or manageable sectors to meet ever-increasing military demands. The structure established distinct unified combatant commands with a dedicated commander and regional duty experts who were jointly responsible for the planning, training, and staffing for contingency operations and real-world scenarios. The unified command structure adopted from the Goldwater-Nichols Department of Defense Reorganization Act of 1986—coupled with an omnibus agreement—further clarified command roles, responsibilities, relationships, acronyms, and terminologies.

Although CentCom was not formed until 1 January 1983, Middle East tensions—including issues stemming from the U.S. hostage crisis in Iran, the war between Iraq and Iran, and the Soviet invasion of Afghanistan—had been simmering for a number of years. Saddam Hussein's aggressive invasion of Kuwait led to Operation Desert Shield in 1990 and, subsequently, Operation Desert Storm in 1991 (together these are known as the Gulf War). The next decade resulted in additional challenging responses that ranged from directing a multinational task force that provided humanitarian relief for the East African nation of Somalia during Operation Restore Hope to monitoring the growing threat of regional terrorists—demonstrated by the 1996 bombing of the Khobar Towers in Saudi Arabia; the 1998 terrorist attacks on U.S. embassies in Kenya and Tanzania; and the October 2000 attack on the USS *Cole* in Aden, Yemen, that killed 17 U.S. sailors.

Given decades of unrest, U.S. government and military strategists recognized the importance of strengthening partnerships in the region to promote economic security and stability. The 11 September 2001 terrorist attacks thrust CentCom into OEF within 30 days, and the operation initially focused on toppling Afghanistan's Taliban regime that was suspected of safeguarding terrorist leader and fugitive Osama bin Laden. Although smaller-scale operations continued in Afghanistan for months, the main military effort shifted to OIF in March 2003.[6]

cil Resolution 1441, which was a final opportunity for Iraq to comply with disarmament obligations, some Security Council members did not endorse such a plan.[7] Regardless of the ongoing political debate in Washington, DC, and at the UN about the validity of a connection between the 9/11 terrorist attacks and Saddam Hussein's weapons program, senior CentCom leaders began planning for a major Iraqi operation just in case.* Army General Tommy Franks directed an intensive eight-month planning cell, incorporating existing Operational Plan (OPLAN) 1003-98 as a foundation. From a facility overlooking Tampa Bay, Florida, at the north point of MacDill Air Force Base, General Frank's nucleus included several planners from OEF who had remained in key joint-staff billets, thus enabling a cohesive, seasoned team to further expand the Global War on Terrorism into Iraq.

After the Gulf War, Saddam Hussein blatantly denied international inspectors on-site access to Iraq's nuclear and biological weapons programs for nearly a decade, which clearly violated UN resolutions and intensified suspicions about his programs and intentions.[8] As a result, in October 2002, Public Law 107-243 authorized the use of military force against Iraq. President Bush endorsed a suspected connection between Saddam Hussein and the 9/11 terrorist attacks, which expanded the search for al-Qaeda terrorists to Iraq. Eliminating Saddam Hussein's dictatorship became a foundation of Bush's presidency, which some of the media along with a handful of vocal Pentagon military and civilian leaders often questioned. Regardless, the president wove this belief into his State of the Union address on 29 January 2003. That evening, he identified Iraq, Iran, and North

*Secretary Rumsfeld pushed his "transformation strategy," favoring lean military campaigns with a focus on special operations, fewer troops, and shorter deployment lead times, which would equate to reduced defense spending and smaller budgets. This was a major shift from the strategy, organization, and defense budgets of the Cold War era when large standing conventional air and land forces were geographically positioned worldwide for combat operations.

U. S. Central Command Key Staff Members, January 2003[9]

Position and Title	Service	Commander	Primary Location
Commander in Chief, Central Command	Army	General Tommy R. Franks	Qatar
Coalition Forces Land Component Commander	Army	Lieutenant General David D. McKiernan	Kuwait
Coalition Forces Air Component Commander	Air Force	Lieutenant General T. Michael Moseley	Saudi Arabia
Commander, U.S. Naval Forces Central Command	Navy	Vice Admiral Charles W. Moore	Bahrain
Commander, U.S. Marine Corps Forces Central Command	Marine Corps	Lieutenant General Earl B. Hailston	Bahrain

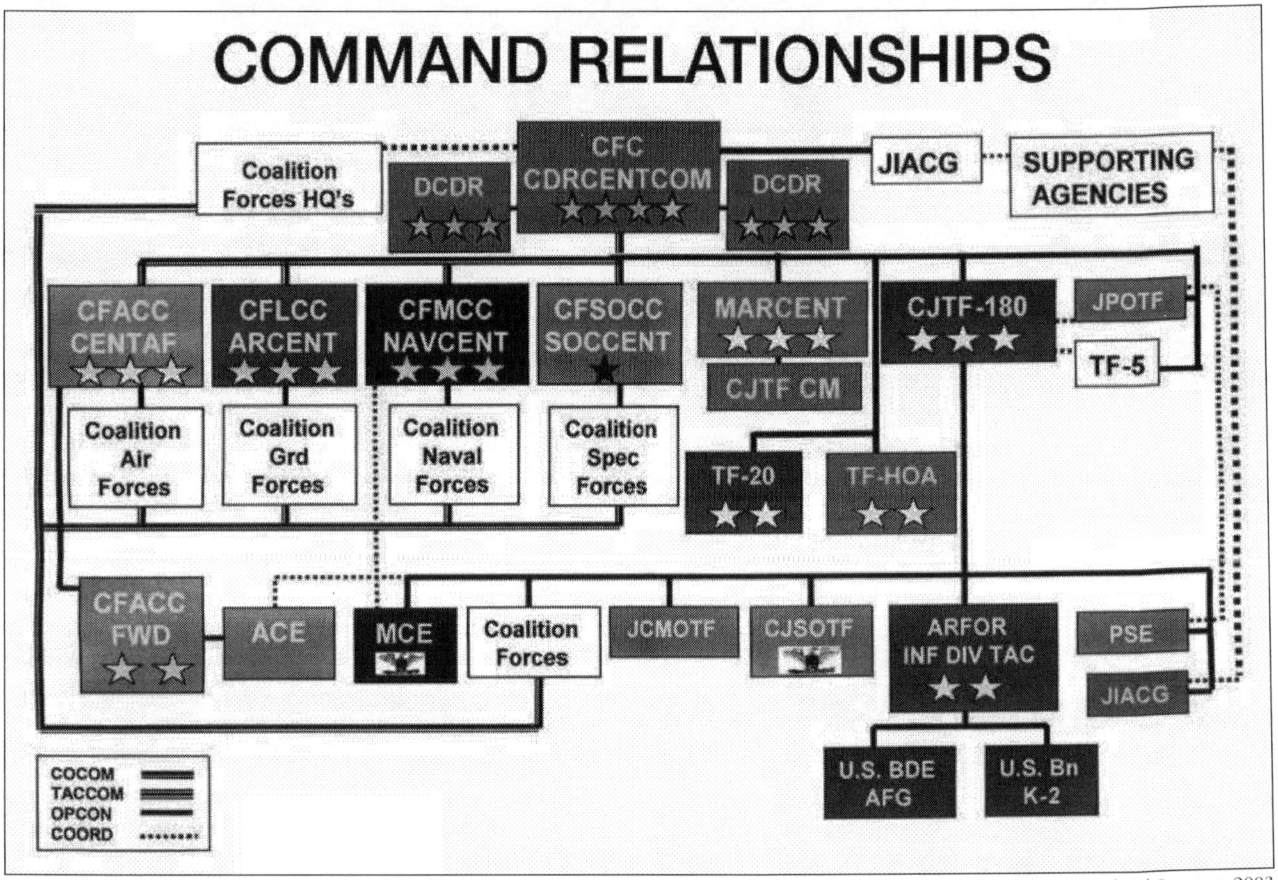

U.S. Central Command, 14 January 2003

Korea as countries that threatened world peace, labeling them the "axis of evil" and clearly setting the stage for a major Iraqi operation.

Realizing that military force might be used to impose UN sanctions against Saddam Hussein's regime to counter the development of weapons of mass destruction, the magnitude of a potential Iraq invasion meant CentCom needed to rekindle Gulf War partnerships and form new ones throughout Southwest Asia. Given cultural differences, business contracts for logistical support most likely would not be negotiated from distant offices or approved quickly. Nonetheless, the command concentrated on the following eight major joint military objectives, as directed by senior-level Department of Defense leadership:

- remove Saddam Hussein's regime;
- identify, locate, and eliminate Iraq's weapons of mass destruction;
- remove al-Qaeda terrorists in Iraq;
- collect intelligence data regarding terrorist networks;
- collect intelligence data regarding global terrorist networks;
- secure Iraq's oil fields and resources;
- end economic sanctions and deliver humanitarian relief to the Iraqi people; and
- assist Iraq in achieving self-government and territorial integrity.[10]

Staffing for a joint service Iraqi operation demanded continual communication and detailed coordination between the service branches. Blending distinct intraservice cultures—operational, organizational, doctrinal, and technical—was commonly branded among military staffs as "purple" operations.* Despite differences, General Franks's commanders set a common tone among their respective staffs that encouraged collaborative relationships throughout the planning and execution phases, particularly among Air Force and Marine Corps aviation commanders.[11]

Until OIF, Marine Corps participation at CentCom headquarters had been characterized as an advisory rather than an operational role. The senior Marine billet had always been headed by a colonel who served as the chief of staff, but without command of operational forces. On 24 October 2002, however, the staffing model changed drastically and expanded to accommodate a surge in activities in the Persian Gulf region, primarily concerning Iraq. Lieutenant General Earl B. Hailston, commander of Marine Corps Forces, Pacific, formally assumed the additional command duties at Marine Corps Forces CentCom, which resulted in numerous flights from his primary base at Camp H. M. Smith in Oahu, Hawaii, to Tampa, Florida, and his forward theater headquarters in Bahrain.[12] The general's newfound role concentrated on fu-

Photo by Amn Bridgett T. Rapp, USAF. Defense Imagery 030320-F-BQ225-044

Commissioned in 1968 through the Enlisted Commissioning Program, LtGen Earl B. Hailston was a career aviator serving in various operational and command billets before assuming multiple roles during Operation Iraqi Freedom, including commander of Marine Corps Forces CentCom. Here, LtGen Hailston (left) meets with troops of the Marine Hercules Aircraft Group, which was colocated with the Air Force's 384th Air Expeditionary Wing.

ture contingencies and identifying resources needed to carry out defined operations. He managed the sustainment effort of OIF, a challenging task, the success of which depended on a continuous flow of equipment, supplies, aircraft, and personnel needed to support air wing operations at multiple sites in Iraq and neighboring countries.

Tailoring a Marine Air-Ground Task Force for Combat

Originally conceived as a proof-of-concept, the Marine Corps tested the staffing model by activating the 1st Provisional Marine Air-Ground Task Force at Kaneohe Bay, Hawaii, on 19 January 1953.[13] In 1986, the Marine air-ground team became a doctrinal cornerstone of Marine Corps operations and remains so today. It consisted of a scalable expeditionary structure that rests on four core interde-

*The Gulf War 12 years earlier introduced and validated joint operations for which Operations Desert Shield and Desert Storm served as prototypes. With expected growing pains, both operations tested a new approach for doing business, and every subsequent operation fine-tuned future joint operations.

Photo by LCpl Kevin C. Quihuis.
Defense Imagery 030213-M-5753Q-001

Standing atop an M1A2 Abrams main battle tank, LtGen James T. Conway, commanding general of I MEF, addresses the 7th Marines regimental combat team at Camp Coyote, Kuwait.

pendent elements, each with unique roles and levels of participation, including a headquarters command element, aviation combat element, ground combat element, and logistics combat element. The MAGTF command element comprises various functions, including intelligence, counterintelligence, ground reconnaissance, communications, information systems, and liaison to the expeditionary force commander, while the other three components provide more direct combined arms and logistical support. Depending on the operational scope and other interrelated factors, planners tailored each MAGTF with its own staffing structure, integrating a balance of combined arms by assigning aviation, artillery, mechanized, and motorized units for appropriate levels of participation.

Given the rugged terrain in Afghanistan, Marine Corps aviation became a major participant in joint operations there, often stretching the MAGTF beyond traditional logistical boundaries. During one particular mission, Sikorsky CH-53E Super Stallion transport helicopters flew more than 350 miles inland from the USS *Peleliu* (LHA 5) to insert Special Forces troops at Forward Operating Base Rhino.[14] Although Lockheed KC-130 Hercules tanker-transports could sustain long-range flights through aerial refueling, operational maneuvers from the sea typically concentrated on littoral areas no farther than 200 miles inland. Just as in Afghanistan, the draft OPLAN 1003V of operations in Iraq outlined a sizeable area, with distances averaging 350–400 miles from permanent air bases in Kuwait to Baghdad. Similar operational and logistical constraints and considerations for sustaining forward air operations would also be encountered. Regardless, the air-ground task organization suited the Marine Corps quite well for this type of operation. Despite its advantages, the force was never designed to sustain air operations at such distances as in Afghanistan or Iraq, yet 3d MAW extended its air control and aviation ground support arm almost 100 miles farther north of Baghdad than planned.

As Pentagon planners provided more details about the scope of the OPLAN, staffing levels and activities at I Marine Expeditionary Force (I MEF) headquarters at Camp Pendleton, California, grew proportionately throughout 2002.[15] The Commandant of the Marine Corps, General James L. Jones Jr., assigned the I MEF commander, Lieutenant General Michael W. Hagee, to lead the planning phase. Hagee simultaneously served as the Corps' primary subordinate unit liaison to the CentCom commanders, General Franks and Lieutenant General David D. McKiernan, USA. One of most appealing attributes of a Marine expeditionary force is its capability to operate with a 60-day self-sustainment package while fighting at the tactical level in support of the joint force commander's campaign.[16]

In the midst of the planning and preparation phase, General Hagee took the oath of office as the 33d Commandant of the Marine Corps, assigning his successor, Lieutenant General James T. Conway, as the commanding general of I MEF.* Without interruption, I MEF formed a doctrinal air-ground team consisting of a headquarters

*Gen Conway, a graduate of Southeast Missouri State University, had just completed a tour as the 1st MarDiv commanding officer, which contributed to a smoother transition as he assumed higher headquarters planning and direction duties.

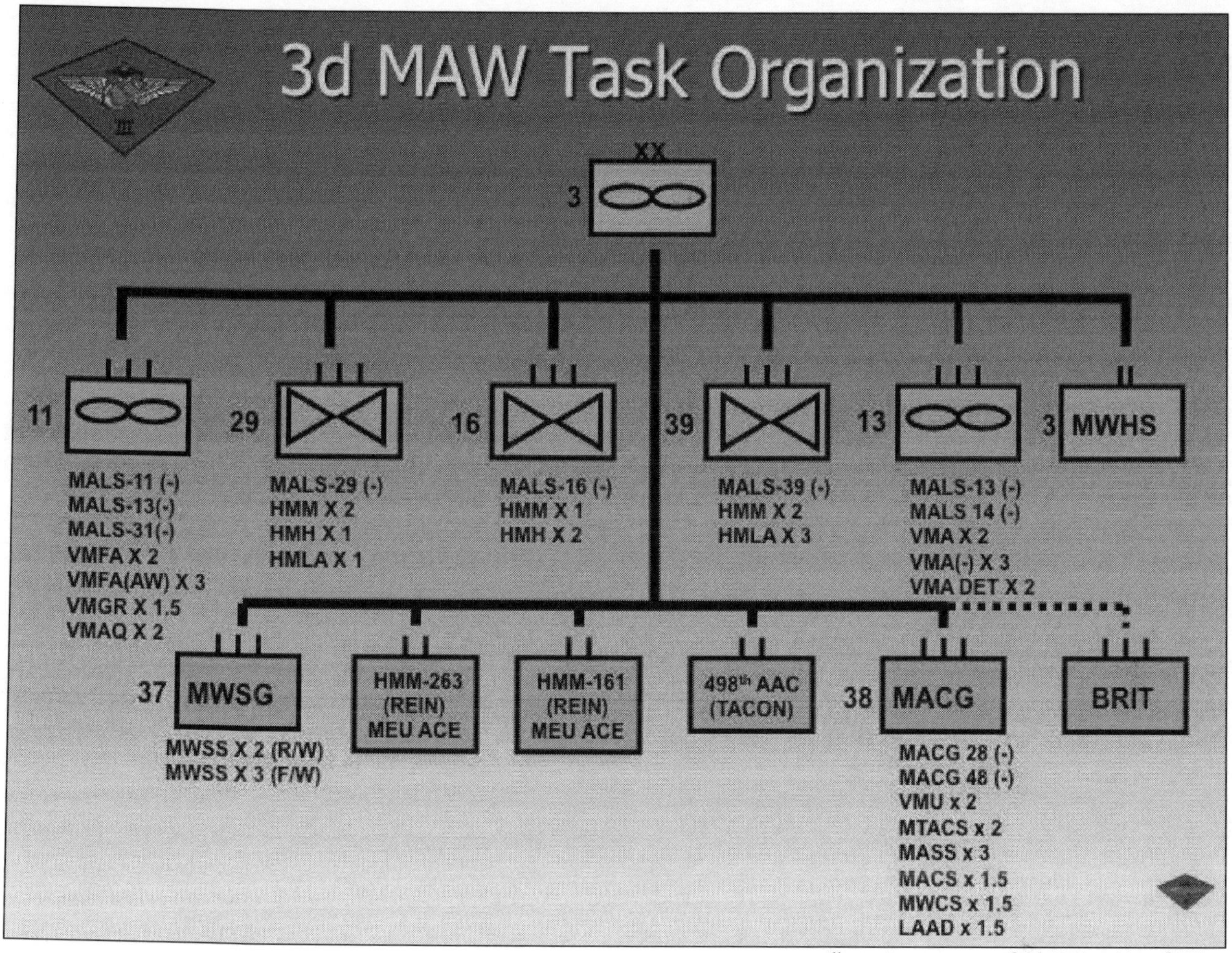

Illustration courtesy of 3d Marine Aircraft Wing

I Marine Expeditionary Force Key Staff Members, January 2003[17]

Unit	Commanding General	Primary OIF Locations
I Marine Expeditionary Force	Lieutenant General James T. Conway	Camp Commando, Kuwait
3d Marine Aircraft Wing	Major General James F. Amos	Ahmed al-Jaber and Ali al-Salem, Kuwait
1st Marine Division	Major General James N. Mattis	Camp Commando, Kuwait
2d Marine Expeditionary Brigade (Task Force Tarawa)	Brigadier General Richard F. Natonski	Camp Commando, Kuwait
1st Force Service Support Group	Brigadier General Edward G. Usher	Camp Commando, Kuwait
1st Armoured Division (United Kingdom)	Major General Robin V. Brims	Camp Commando, Kuwait

command element and three additional components: an aviation combat element from 3d MAW, a ground support element from 1st MarDiv, and a combat service support element from 1st Force Service Support Group (1st FSSG).[18]

As the operational scope expanded, other units augmented I MEF and eventually included Task Force Tarawa ground units from the 2d Marine Expeditionary Brigade (2d MEB) based in Camp Lejeune, North Carolina, reinforced with 2d Marine Aircraft Wing (2d MAW) squadrons that eventually joined Marine aircraft groups; the 15th Marine Expeditionary Unit (Special Operations Capable) (15th MEU [SOC]); the 24th Marine Expeditionary Unit (Special Operations Capable) (24th MEU [SOC]); the command element of the 11th Marines; the I MEF Engineer Group; medical and air defense units from the U.S. Army; and the 1st Armoured Division from the United Kingdom. When combat operations began in March 2003, I MEF expanded to include more than 86,000 Marines, sailors, soldiers, and British troops that were primarily assigned to bases in Kuwait, Bahrain, and Saudi Arabia.[19]

Generals Conway, Amos, and James N. Mattis were not strangers before assuming their respective commands during OIF. As senior leaders, they had moved to the highest echelons of Marine Corps command, often crossing paths during operational exercises, professional military education, and official events. Although the officers had varying operational backgrounds and leadership styles, they formed an unusually strong bond that set an example of collaboration and teamwork for their respective staffs. Because of their first names, fellow Marines often referred to them as the "three Jims."

Forming an Aviation Combat Element

In August 2002, General Amos had wrapped up his final Pentagon tour as assistant deputy commandant for Plans, Policies, and Operations, Headquarters Marine Corps, to take command in dual roles—as commanding general of 3d MAW and as commander of the aviation combat element. Amos was well prepared for these roles with a resumé that balanced flight hours, group command, staff tours, and joint operations. A career aviator, the general had earned his naval aviator wings and initially flew the McDonnell Douglas F-4 Phantom II before transitioning to the F/A-18 Hornet. In addition to operational assignments, he had held various command and joint staff billets, most notably serving as the chief of staff of U.S. Joint Task Force Noble Anvil during the air campaign over Kosovo, which exposed him to both the benefits and challenges of joint air operations.

General Amos's aviation combat element integrated aircraft, aviation assets, and personnel from three air wings, forming a command headquarters, five aircraft groups, an air control group, and a wing support group. Specifically, 3d MAW deployed helicopters and personnel from its organic California-based groups: Marine Aircraft Group 16 (MAG-16) at Marine Corps Air Station Miramar and MAG-39 at Marine Corps Base Camp Pendleton. Fighter and attack aircraft and personnel came from F/A-18 Hornet squadrons within MAG-11 at Miramar and from McDonnell Douglas AV-8B Harriers assigned to MAG-13 at Marine Corps Air Station Yuma, Arizona. The 2d MAW reinforcements came primarily from Cherry Point and New River, North Carolina, and Beaufort, South Carolina, adding MAG-29 and assigning squadrons to other aircraft groups. Support personnel from Marine Air Control Group 38 (MACG-38) and Marine Wing Support Group 37 (MWSG-37) that were headquartered at Miramar and San Diego, California, provided critical air control and sustainment support, both integral components within the aviation combat element. In total, 3d MAW oversaw five flying aircraft groups along with MWSG-37 and MACG-38. MACG-38 also had air control elements located at Camp Pendleton; at Yuma; and at the Marine Corps Air Ground Combat Center in Twentynine Palms, California.

Not surprisingly, wing planners immediately faced enormous logistical challenges in determining suitable locations for launching sustained air operations in the Gulf region where prime real estate was already limited. Another pressing concern centered on the oversight and ownership of aviation assets in the joint airspace—an unresolved, lingering issue from the Gulf War and subsequent joint operations. The Marine Corps' expeditionary nature demanded an immediate tactical air request process to adequately support its ground forces. Additional layers of tactical decision making or approval in joint operations would hamper responsiveness. Marine pilots, air controllers, and ground commanders therefore needed a timely scheduling and assignment process that would facilitate immediate missions yet be within compliance of the U.S. Air Force's 96-hour air tasking order cycle—i.e., the master schedule—for assigning targets, aircraft, and missions. Marine aviation had always taken great pride in its air command-and-control system, so efforts to retain centralized command and decentralized control of its aircraft during Iraqi operations was a top priority for General Amos and his air planners. The aviation combat element had to balance planning and providing expeditionary air support for Marines on the ground, while providing aviation resources and operating under an Air Force umbrella.

The Moseley-Amos Agreement

Although not a new concept, joint air operations were often strained, given unique service cultures and differences in mission types and command-and-control procedures; however, OIF afforded Marine aviation an opportunity to implement a fresh way of tasking its aircraft. The most significant command relationship that influenced how Marine aircraft were assigned was between U.S. Air Force Lieutenant General T. Michael Moseley, the Coalition forces air component commander, and General Amos. Fully anticipating a heated debate over command of the airspace and control of hundreds of aircraft, General Hailston arranged an introductory meeting among key players early in October 2002 to hash out the details. The intent was to spark face-to-face dialogue to determine how Marine aviation assets would be managed and employed within the joint environment. Typically, the Air Force focused on planned strategic aviation missions with a longer planning and targeting cycle, while the Marine Corps' expeditionary nature required more immediate or on-call fixed- and rotary-wing support for Marines on the ground. As a result, Service philosophies were seldom in alignment. During the Gulf War, elements of Marine aviation were managed by joint planners, which created an extra layer between Marine pilots and the troops on the ground, thus negatively impacting response times.

General Hailston realized that the commanders who would oversee the joint air war needed to begin discussions regarding air command and control, so he introduced Generals Moseley and Amos—who had never met—to jump-start conversations and set the stage for a positive working relationship. Already serving as the senior air commander in the Iraqi theater and enforcing Iraqi no-fly zones by overseeing Operations Northern and Southern Watch, Moseley and his staff had established a sound airspace management plan, and the Marines would need to comply with existing protocols. Amos was a firm believer in teamwork and relationship building, and he was prepped to highlight his position that 3d MAW would add an enormous capability, deploying with more than 400 aircraft, but he wanted the Marines to control their own airplanes.

Photo by MSgt Jim Varhegyi, USAF. Defense Imagery 040225-F-3050V-105

U.S. Air Force Gen T. Michael Moseley's understanding of the Marine air-ground task force facilitated a positive working relationship in managing joint aviation assets during Operation Iraqi Freedom.

General Moseley arrived for the Friday afternoon meeting in October 2002 in civilian attire and was greeted by Marine Corps Lieutenant Generals Conway and Hailston, Major Generals Amos and Mattis, and Brigadier General Edward G. Usher III. Surprisingly, the discussion did not result in an argument as many might have expected, but rather it evolved into a meeting of the minds. As General Amos recalled, General Moseley defused any tension with a few powerful words: "Let me tell you guys something. First of all, I am a big fan of Marine aviation; I am a big fan of the Marine air-ground task force. General Hailston and I are friends, and we go back to our days at the National War College. I want Marine airplanes flying over Marines. So I'm okay with Marine airplanes supporting Marines. I'm feeling good about this."

Moseley looked at each of them and then continued, "I own every millimeter of airspace above the ground and up, and I want to make sure everybody understands that there's not going to be some Marine bubble." Both air commanders agreed to let their staff resolve details, such as how to list missions on the air tasking order. Not long after the meeting, Marine Colonel Ronnell R. McFarland, commanding officer of MACG-38, began logging frequent flier miles because of meetings with Air Force planners at Shaw Air Force Base, South Carolina, where they further defined a joint airspace plan.[20]

The Marines did not need to provide a series of convincing arguments because General Moseley declared himself a MAGTF advocate and displayed an accurate understanding of Marine Corps–style centralized air command and decentralized air control. Fully grasping the Corps' aviation combat element role, he announced that General Amos could task his air assets in direct support of the Marines, but with one nonnegotiable condition—ownership or supervision of the air war would remain with the joint forces air component commander. In essence, 3d MAW aviators would support Marines, but excess sorties would be available for joint assignment tasking.

The single exception to this agreement was oversight of Marine Corps Grumman EA-6B Prowler electronic warfare aircraft. Considered a national asset by the Joint Chiefs of Staff, the few Prowlers available would remain under Air Force control and assignment in the joint arena. While Navy Prowlers operated from carriers, 10 Marine Corps aircraft flew from the Air Force's combined air operations center at Prince Sultan Air Base, Saudi Arabia. Understanding General Moseley's rationale, yet unsure of its application during the air campaign, the Marine Prowler community had mixed reactions about its segregation from the Marine aviation combat element, especially since the aircraft's jamming capability supported one of the six Marine aviation functions: electronic warfare. With no direct oversight, 3d MAW would be forced to forward unit-initiated EA-6B air requests to the combined air operations center—not directly to Marine Corps control agencies—adding what seemed like an extra approval layer along with precious minutes, even hours, to response times.[21]

Nevertheless, the Marine Corps complied, fully understanding that its Prowlers would play a huge role in disrupting enemy communications and conducting psychological operations, thus benefiting all Coalition forces. As the 3d MAW commander summarized, the "Moseley-Amos agreement" was a hallmark milestone, a genuine "grand slam for Marine aviation," altering the way joint air operations had been executed in the past. In the end, giving up excess sorties during OIF was an agreeable tradeoff that the Marine Corps readily accepted.[22]

Chapter 2
Understanding the Ground Scheme of Maneuver

Anticipating Hostilities

Before 3d MAW planners could develop a viable air campaign, they needed to fully understand the ground scheme of maneuver. The ground strategy was the pivotal component within the MAGTF that affected how the air wing would deploy its flying squadrons and would support fixed-base and forward air operations. Assessments of current Iraqi and Kuwaiti air base conditions and runway lengths along with determining the appropriate aircraft mix, spare parts and support equipment inventories, ordnance configurations, and proposed locations of air command-and-control sites were key interrelated components in building the air strategy to complement the ground advance. Answers to these fundamental questions evolved over time and provided the foundation that further defined critical aviation-specific operational, logistical, and augmentation requirements. Given the nature and complexity of air operations, air-ground team planners worked on solutions months in advance of the deployment.

General Mattis, fresh from leading Task Force 58 operations and directing Marine participation in Afghanistan during OEF, assumed command of 1st MarDiv at Camp Pendleton, California, in August 2002 in preparation for OIF. A few hours after the change-of-command ceremony, he summoned the division's senior leadership for their first staff meeting. Unit commanders, principle staff officers, and sergeants major assembled in the conference room expecting to hear the new commander's message of intent. General Mattis was regarded by those who had worked for him as a dynamic leader who demanded first-class performance and molded his staff into a cohesive team-oriented unit that could respond quickly and decisively. Typically, he provided commanders with guidance, but he also strongly encouraged the staff to solicit new ideas and to think creatively.[1] The meeting was unlike the social gathering that typically followed a change-of-command ceremony; it was businesslike and characterized by a serious tone. The general cordially welcomed the incoming staff and then issued an informal warning order for an Iraqi offensive operation.[2] He presented a vision for a full-scale war that centered on staff planning; unit readiness; predeployment training; and, once in combat, decisive maneuvers coupled with speed. His introductory words set the stage for an eight-month intensive planning effort that did not end until OIF began on 20 March 2003.[3]

Photo by SSgt Daniel E. Schrubb.
Defense Imagery 030201-M-2081S-028

MajGen James N. Mattis, commanding general of 1st MarDiv during Operation Iraqi Freedom, addresses Marines of Headquarters Battalion after a promotion ceremony at Camp Commando, Kuwait.

Generals Amos and Mattis firmly believed that their staffs had to form a cemented partnership to gain an in-depth common understanding not only of their respective roles and missions, but also of the country where they would probably confront Islamic extremists and terrorists. Iraq had been a familiar coordinate on command maps since 1st MarDiv participated in several joint training exercises and operations on the Arabian Peninsula after the 1990–91 Gulf War. Military action and victory in Operation Desert Storm had familiarized a new generation of Marines with the Middle East when the United States and an international coalition thwarted Saddam Hussein's attempt to expand into neighboring Kuwait.[4] Although the Gulf War had temporarily contained Hussein, the dictator's repeated refusals for more than a decade to cooperate with UN weapons inspectors were gaining worldwide attention and generating impatience.

Generals Amos and Mattis formed an air-ground team and filled key billets with Marines from squadrons, aircraft groups, battalions, and regiments. Many of these Marines had participated in the Gulf War, as well as in the Somalia, Bosnia, and Kosovo campaigns. Operation Desert Storm veterans who had been promoted during the past decade to senior officer and noncommissioned officer ranks shared their institutional knowledge and personal experiences regarding air operations in an austere, desert environment. The 12-year gap between major Iraqi operations meant that senior officers and staff noncommissioned officers provided a valuable perspective on Saddam's military forces for a much younger generation of Marines who had been elementary and middle school students in 1991. General Mattis realized that the corporals and captains who would most likely confront the enemy face-to-face during ground hostilities needed to understand the cultural, political, and economic context of the Middle East that included not only "axis of evil" members—Iraq and Iran—but also Syria, an acknowledged sponsor of terror.

The 1st MarDiv intelligence director, Col James R. Howcroft, assigned three lance corporals to create and distribute an orientation map of southern Iraq that would familiarize units with a common reference of prominent topographical features.

Map courtesy of 1st Marine Division

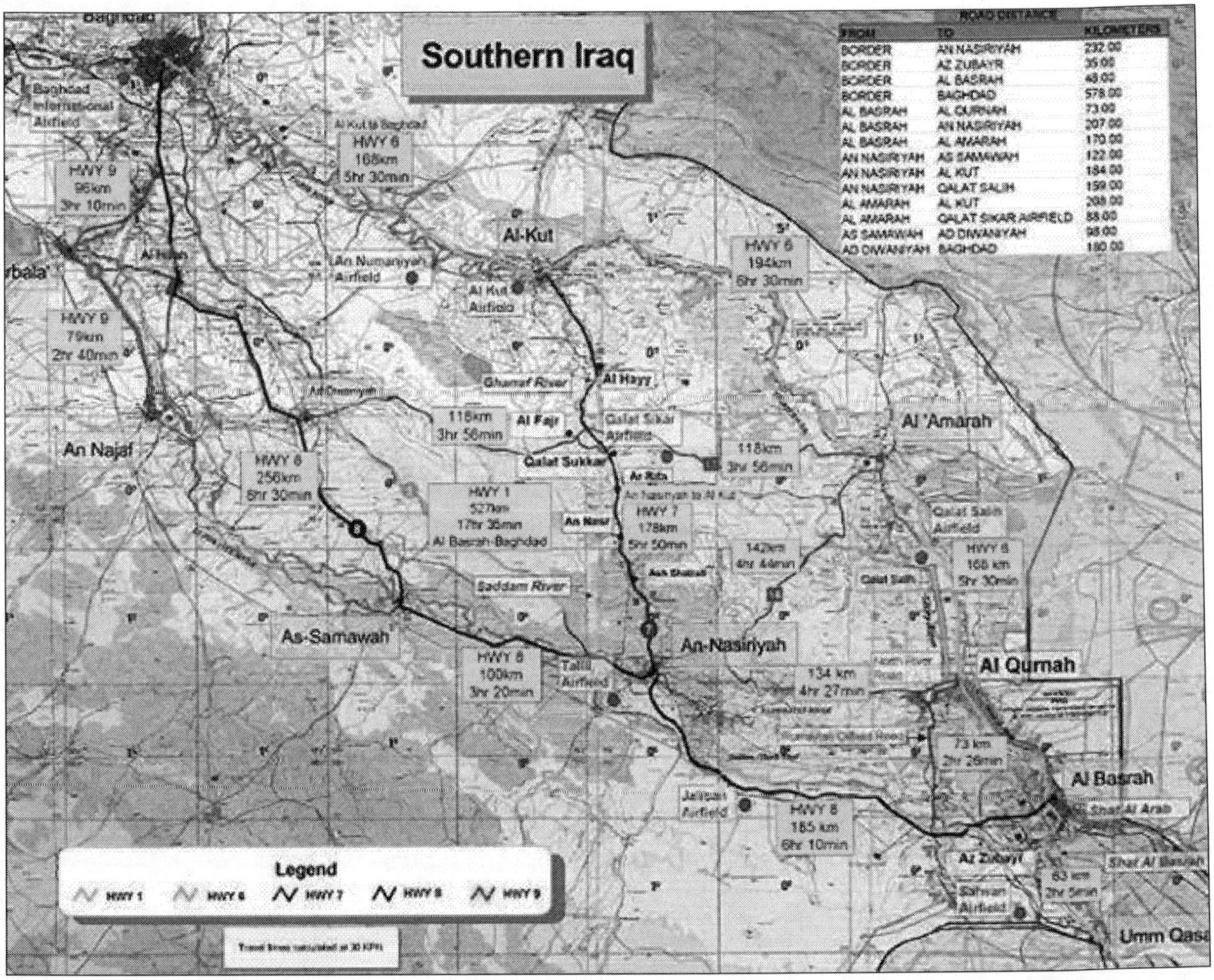

Iraqi Heritage

Iraq was home to a long line of ancient Mesopotamian civilizations, including Sumer, Akkad, Assyria, and Babylonia. The region fell to Cyrus the Great of Persia (in the sixth century BC), Alexander the Great (in the fourth century BC), the Arabs (in the seventh century AD), and the Ottoman Turks (in the 16th century AD). Established as an independent kingdom in 1921, Iraq became a republic after the assassination of King Faisal II in 1958.[5]

At the time of OIF, the Iraqi population consisted of three distinct groups: the Shia majority; the ruling Sunni minority; and the Kurds, who have lived a more-or-less autonomous existence in the northeast corner of the country with large, well-armed militias. Neither the Kurds nor the Shia shared the Sunni admiration for Saddam Hussein, who assumed power in 1979.[6] The Sunni population includes both Arabs and Kurds. Most Kurds are Sunni Muslims, but they differ from their Arab neighbors in language, dress, and customs. While Arabic is the most commonly spoken language in Iraq, Kurdish is spoken in the north, and English is the most prevalent Western language.

Understanding Iraq—Then and Now

Long before Operation Desert Storm or Operation Iraqi Freedom began, the Iraqi people had endured a volatile, economic roller coaster ride that was attributed to deteriorating regional relations and destructive decisions by the Saddam regime. In September 1980, the outbreak of the Iran-Iraq War brought Iraq's once profitable oil production to a halt after peaking at an estimated $35 billion in foreign exchange reserves before the war. The loss of oil exports had an immediate negative economic impact that directly affected the nation's future growth and stability. Ultimately, the Iraqi government was saddled with a foreign debt of more than $40 billion. Although oil export production gradually increased with new pipeline construction, Saddam caused another economic tailspin when he ordered the Iraqi invasion of Kuwait that precipitated the Gulf War.

After the Gulf War, Saddam Hussein aggressively tried to suppress the Kurdish population in northern Iraq and the Shiite population in the south. As a result, the UN Security Council passed Resolution 688, which condemned Baghdad's repression of its Shiite and Kurdish populations. In 1992, the United States, Great Britain, and France established two no-fly zones over the country—one extending north of the 36th parallel and one extending south of the 32d parallel (expanded to the 33d in 1996)—to protect the two populations. The containment of Saddam's airpower further degraded Iraqi pilot training and skills.[7] Although never sanctioned by the UN, the monitoring and control of these no-fly zones—operations named Northern and Southern Watch—clearly defined restricted Iraqi airspace and limited its airpower, thus ensuring Coalition air superiority.

In December 1996, the UN's oil-for-food program offered a glimmer of hope to the Iraqi people and allowed Iraq to export limited amounts of oil to finance essential civilian needs, including food, medicine, and infrastructure repair. Despite these positive measures, years of the dictator's brutality had taken their toll on the spirit of the people, constraining future economic, educational, and social enrichment. Moreover, further constraints resulted from Saddam Hussein's regime itself, which was comprised of 18 unique governorates with minimal authority for decentralized decision making or oversight. As the economy faltered even more, hope and the quality of life for average Iraqis plummeted while Saddam's army and Baath Party loyalists flourished, leading lavish, opulent lives.

The country struggled to provide the essential services—electricity, water, transportation, telecommunications, and oil—required for a stable economy. Security and educational services were also woefully inadequate. The electric supply, in particular, was erratic and unpredictable. While residents of Baghdad could expect between 12 and 24 hours of service each day, only 4–8 hours of power were available for the average citizen outside of the capital. Even more pressing, only 5.5 million of Iraq's 25 million citizens had ac-

Photo by Sgt Paul L. Anstine II. Defense Imagery 030402-M-5150A-026

A destroyed Iraqi T-55 tank burns near the an-Numaniyah Bridge along Highway 27 during Operation Iraqi Freedom.

cess to a safe and stable water supply prior to March 2003. Cities suffered from inadequate sewage systems and ineffective, ethnic-based health care. Border security control measures languished because of a weakened security force. Most problematic for Iraq's future was the fact that although the Ministry of Higher Education and Scientific Research consisted of 22 universities and 46 institutes or colleges within the community college system, only 2.96 million (less than 50 percent) of the approximately 6.1 million children who were enrolled in Iraq's lower education system were expected to graduate from secondary school.[8]

Current Enemy Military Threat

The U.S. State Department published military assessments after the Gulf War indicating that Saddam Hussein's army had shrunk to 23 divisions (from a high of about 68 divisions) and the air force was reduced to less than 350 aircraft (from more than 500).[9] Reports of inadequate troop strength, unit capabilities, and parts and supplies availability further highlighted Iraq's weaknesses and vulnerabilities. Despite the war's impact on his military forces, Saddam still devoted enormous efforts toward purging any hints of opposition and dissent. He aggressively tried to suppress the Kurdish population in northern Iraq and directed major division-level counterinsurgency operations northwest of Basrah by approving large-scale engineering efforts to divert water away from marshlands to facilitate combat operations. Many Iraqis—most notably Shiites in the south—rebelled against his government, which responded quickly and with crushing force, killing thousands. Even after years of containment and economic sanctions, Saddam retained a smaller yet significant conventional combined-arms capability that warranted keen interest by the 1st MarDiv, particularly regarding indirect fire capabilities. In March 2003, the Iraqi military threat comprised four primary factions: conventional forces of the regular army and the *Republican Guard*; special forces units; Baath Party militia; and Saddam's loyalists, the paramilitary *Saddam Fedayeen*.

One of General Mattis's major concerns, aside from indirect fires, was the potential for Saddam's military to unleash weapons of mass destruction. It was known that the dictator had used chemical weapons against the Kurdish population in northern Iraq.[10] Intelligence analysts reported that, even though the Iraqi army was still organized, it had been reduced to a third of its size since the Gulf War and was poorly trained, equipped, and managed. Regardless of lower troop strength, Saddam's regime had a reputation for brutal, centralized command and control that demanded strict obedience from subordinate commanders; they were ruled by fear, and initiative and small unit leadership were not encouraged. Strategically, if communications were disrupted in Saddam's core headquarters, his chain of command would quickly break down, affording U.S. planners an opportunity to exploit key weaknesses in the heart of Baghdad.[11] Despite a weakened conventional military profile, and perhaps because of it, the potential for using weapons of mass destruction remained an unsettling prospect.

Army and Marine Corps Set Roles and Boundaries

The gist of OPLAN 1003V entailed joining forces from the V Corps—"Victory" Corps—of the U.S. European Command, led by U.S. Army Lieutenant General William S. Wallace, and the Marine Corps' I MEF, led by Lieutenant General Conway. Both units would cross into Iraq from the northern border of Kuwait on a two-front attack to Baghdad, using separate avenues of approach. The 3d Infantry Division, led by Major General Buford C. Blount III, USA, would advance to Baghdad on a course south of the Euphrates River, securing the Highway 1 bridge near the outskirts of an-Nasiriyah before continuing their charge northwestward through the open desert. Meanwhile, 1st MarDiv would race north to Baghdad from the east, playing a critical supporting role. Identified as the enemy's center of gravity, Baghdad was where the bulk of Iraqi military forces were suspected to be positioned.

Based on Army and Marine Corps projections, 1st MarDiv was slated to reach the outskirts of Baghdad within 55 days after crossing the line of departure. Speed was the cornerstone of the ground strategy, and Iraq's varied landscape and harsh desert climate presented significant obstacles. Bordered by six countries—Turkey, Iran, Kuwait, Saudi Arabia, Jordan, and Syria—Iraq consists of four distinct zones or regions: the sparsely

A pair of captured Russian-designed Iraqi FROG-7 surface-to-surface artillery rockets rest on Republican Guard *launch trucks along Highway 8 near ad-Diwaniyah.*

Photo by Sgt Paul L. Anstine II. Defense Imagery 030424-M-5150A-013

populated desert area that accounts for 38 percent of the country's geography in the west and southwest; the rolling upland region between the Tigris and Euphrates Rivers; the mountainous highlands in the northeast on the northern Iranian and Turkish borders; and the central and southeastern alluvial plain, characterized by a low, marshy wetland that begins north of Baghdad and stretches south to the Persian Gulf. Of most concern was overflow water from canals and lakes, which created the potential for natural flooding and swampy terrain from Iraq's two main rivers—the Euphrates and Tigris.

As the previous I MEF commanding general, Lieutenant General Hagee's guidance noted that it would be better to fight harsh terrain than attack pockets of enemy resistance. General Mattis's leadership team designed a plan that sustained forward momentum by avoiding engagements with the enemy in defensive positions along their route.[12] The core strategy demanded a rapid march north by crossing over the Euphrates and Tigris Rivers, bypassing urban areas where suspected Iraqi units might be waiting. Given geographical constraints, the fastest means to Baghdad meant the war would not be fought in the open desert, but rather alongside secondary roads with a much-needed logistical train of tactical convoys stretching from Kuwait to Baghdad. The journey from northern Kuwait to Baghdad, and subsequently to Tikrit, totaled almost 450 miles. In comparison, the distance from Kuwait City to Baghdad was nearly the same distance as from Los Angeles to San Francisco, about 350 miles, while the distance from Baghdad to Tikrit added another 85 miles. Although 1st MarDiv planned to bypass urban areas with significant concentrations of civilians, troop movements faced other obstacles along the route while fording water crossings and encountering inadequate highways compared with American standards. As it turned out, the rapid pace created vulnerable logistical convoy and channelized supply lines, which depended even more on Marine aviation for a host of aerial reconnaissance, aerial resupply, and close air support missions.

If executed as planned, with the Army advancing undetected west of Baghdad through the open desert and the Marines traveling east of the Euphrates River, the probability of major enemy engagements was reduced. By avoiding large populated areas, the intent was to surprise enemy military forces, quickly seize control of Baghdad, and stifle Saddam's power and influence at his center of gravity, therefore minimizing the likelihood of massive civilian or Coalition casualties.[13] In essence, the Marines were creating a diversion for the Army by drawing enemy attention away from Baghdad by making the Marines the enemy's focal point on the eastern flank.

The success of the final phase—security and

Boundaries Change

On 9 February 2003, 1st MarDiv benefited from a huge change in battlefield geometry, resulting in a real-world use of the Marine Corps planning process. The unfinished Highway 1 corridor between an-Nasiriyah and Baghdad, just west of the 1st MarDiv zone, had previously been in the Army's area of operations. During initial planning in summer 2002, the Marines had tried to gain access to this area for their envelopment of al-Kut, but the Army would not change boundaries. However, less than five weeks before hostilities started, the Army changed its avenue of approach to Baghdad and shifted the bulk of its combat power to the western attack corridors of its planned zone.

Subsequently, the area north of the Euphrates River, or the eastern flank, would not be used, so Marine Corps leadership quickly volunteered to take ownership of this terrain. Division tactical planners recognized that this route would significantly improve speed because it offered a second path rather than just one highway to Baghdad. As an additional bonus, the road was only partially finished, so Saddam's forces might not expect a major advance north of Highway 1, which afforded a golden opportunity to achieve tactical surprise. The ability of the Marine Corps to adapt at such a late date before combat operations began once again demonstrated the flexibility and benefits of the MAGTF model.[14]

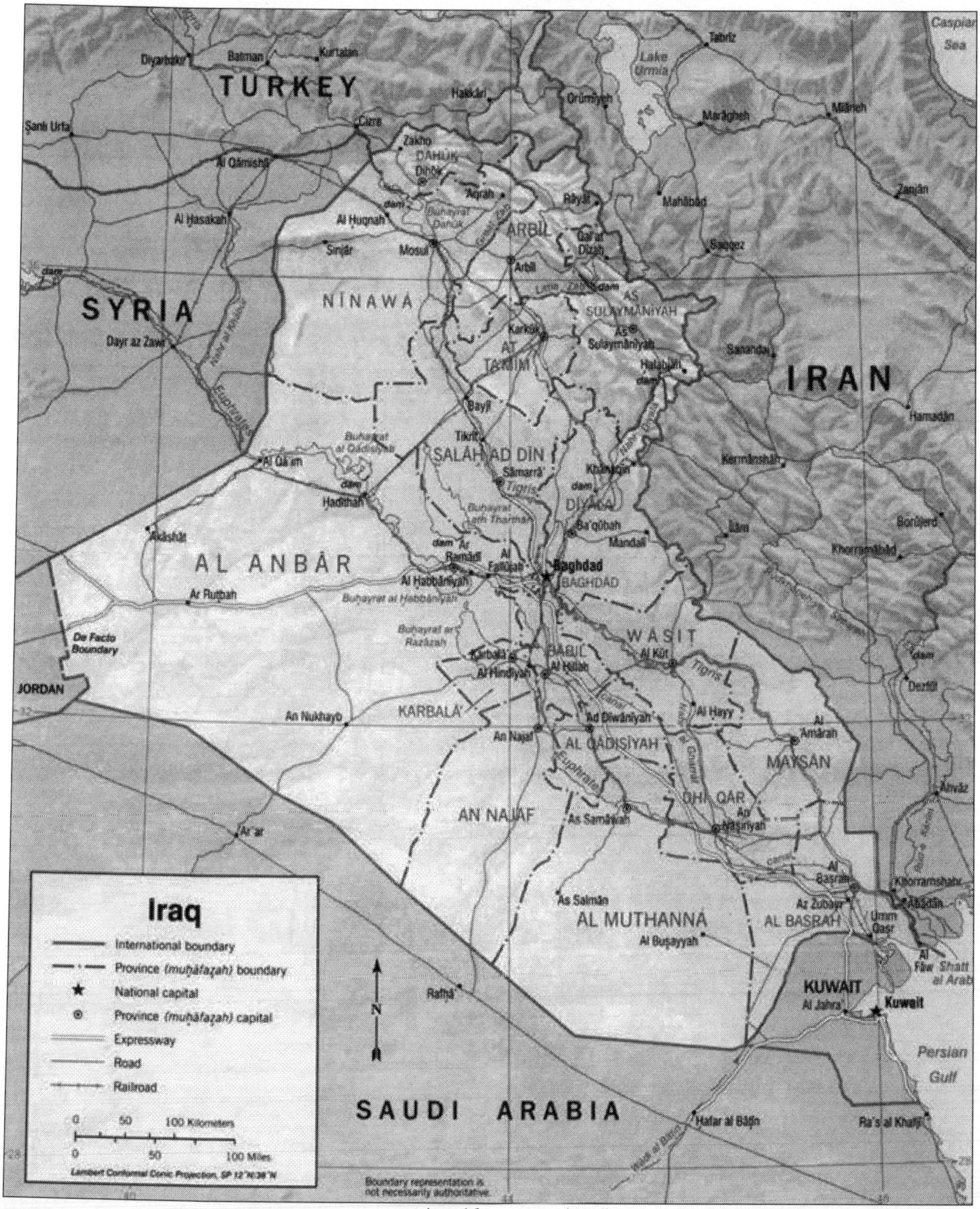

Adapted from a Central Intelligence Agency map by Marine Corps History Division

stability operations—depended on three key prerequisite assumptions that undoubtedly influenced the Marine air-ground strategy. First, U.S. military rules of engagement stressed minimal collateral damage during the decisive combat phase because a major rebuilding effort was neither anticipated nor planned in detail. Accordingly, 3d MAW crews would need to fly precise bombing

The Four Phases of Operation Iraqi Freedom

Phase	Scope
I Preparation	Plan the logistical movement of personnel, aircraft, equipment, and supplies, and solicit regional and international Coalition support
II Shape the battlefield	Prepare the battlefield before ground operations
III Decisive operations	Coordinate offensive air-ground operations to defeat Iraqi forces and remove Saddam from power
IV Post-hostilities operations	Provide security and stability after combat operations

missions, taking full advantage of the wing's mix of guided munitions. Second, U.S. military strategy avoided offensive battles near urban areas (other than Baghdad) to sustain their momentum and minimize enemy diversions from the main effort. Third, after Saddam's regime fell, planners expected that oil output would suffer only temporary setbacks and would provide a valuable revenue stream to fund rebuilding efforts. Given these assumptions, phase IV detailed planning and coordination among government agencies and military commands had been minimal.

Speed and the Need for Marine Aviation

Anticipating a projected troop strength of 80,000 personnel, I MEF commanders determined early on in the planning for OPLAN 1003V that maneuver warfare—coupled with speed and surprise—formed the critical path for the Corps' success. As defined by Marine Corps doctrine, "Maneuver warfare is a warfighting philosophy that seeks to shatter the enemy's cohesion through a variety of rapid, focused, and unexpected actions which create a turbulent and rapidly deteriorating situation with which the enemy cannot cope."[15]

Understandably, General Mattis was most concerned with minimizing the enemy threat and destroying Iraq's indirect fire weapons—artillery, surface-to-surface rockets, and launchers—especially as ground forces approached Baghdad. Targeting focused on Iraqi systems capable of firing chemical and conventional weapons at ranges that exceeded those of Marine artillery batteries. This required reliance on air support, thus leveraging the integrated air-ground team to minimize this enemy threat.[16] Learning even more about the dictator's brutality toward his citizens as the confrontation loomed, General Mattis repeatedly reminded Marines that hostilities were aimed at the Saddam's military regime—not the Iraqi people. The division commander developed a "speed equals success" strategy that would prove its effectiveness by paralyzing the hub of enemy communications and destroying their will to fight, yet balanced it by minimizing the loss of Marine and innocent Iraqi civilian lives. General Mattis conveyed this message again during his commander's intent to Marines and sailors two days before the air campaign began, reemphasizing the division motto: "Demonstrate to the world there is no greater friend, no worse enemy, than a U.S. Marine."[17]

Given this strategy, MAGTF planners realized that the means to effectively achieve maneuver warfare rested in integrating its air assets to disrupt Iraq's command-and-control capabilities as the 1st MarDiv pushed toward Baghdad. Ground commanders would depend on Marine aviation capabilities to provide not only logistical and air support, but also intelligence and firepower well ahead of the division and protecting its eastern flank. The wing planners anticipated a mix of aerial troop insertions, armed escorts, and close air support missions along with air support to disrupt enemy activity and destroy command-and-control sites deep inside Iraq.[18] These missions meant that Marine aviation would also have to orchestrate a robust logistical support system, operating from permanent base operations in Kuwait and small, temporary sites inside Iraqi territory along the dual routes toward Baghdad.

Defense Imagery 030324-M-9124R-014

Reinforcing Gen Mattis's strategy, an AH-1 Cobra helicopter crew scans for enemy activity above Company D, 1st LAR, during Operation Iraqi Freedom. The 3d MAW planned for both close and deep air support missions while responding to immediate on-call requests during 1st MarDiv's march to Baghdad.

In turn, these requirements generated a host of considerations for planners, such as determining suitable locations for remote air operations, aircraft disposition, critical resupply points, and intermediate-level maintenance activities. Of grave concern was the replenishment of aircraft fuel and munitions at remote airfields so that fixed- and rotary-wing crews could extend their time on-station closer to the Marines on the ground. Although aerial refueling for certain aircraft would alleviate part of the strain, not all aircraft had that capability. Other interrelated "million-dollar" questions emerged as 3d MAW planned the air campaign for a massive deployment to Kuwait and, subsequently, hostilities deep inside Iraq.

Chapter 3

Developing the Marine Aviation Strategy

Differences from the Gulf War

During the Gulf War, more than 500,000 Coalition combatants fought against Saddam's invasion of Kuwait, destroying elements of the dictator's *Republican Guard* assembled in southern Iraq.[1] On 17 January 1991, the Coalition decisively gained air superiority on the first evening of the air campaign and continued preplanned air strikes for six weeks. Twelve days after the air war began, the Iraqis launched an attack into Saudi Arabia, the first major ground combat action of the Gulf War, at the town of al-Khafji.[2] In this battle, joint air strikes demonstrated the capabilities of air-ground team operations and served as a prototype for the ground war nearly a month later.

Unlike the strategic air campaign that dominated Operation Desert Storm, CentCom's plan for orchestrating the 2003 Iraqi invasion was quite different. OPLAN 1003V depended on tactical surprise through the rapid advancement of troops to Baghdad.[3] To accommodate this, General Amos built a strategy that leveraged four key advantages during the march, each of which highlighted the strengths of the aviation combat element.

The first advantage was U.S. air dominance over Iraq. This had been gained even before OIF launched its first mission, or sortie;* therefore, a lengthy period of preliminary air strikes or battle shaping before ground options was unnecessary. Iraqi air power had been severely blunted by more than a decade of missions—some 300,000 sorties—flown by Coalition pilots after the Gulf War in support of Operations Northern and Southern Watch.[4] In 2003, the Iraqi integrated air defense system comprised approximately 150 early warning radars and fewer than 350 fighter-attack aircraft. For the most part, these metrics painted a relatively low enemy air threat. However, U.S. Air Force analyses reported 200 surface-to-air missiles concentrated mostly around Baghdad.[5] In an attempt to neutralize the missile threat, CentCom increased the tempo of bombing missions just weeks before hostilities began—using Operation Southern Focus as a catalyst. This operation, which was not officially announced until mid-2003, began in June 2002 and was designed to counter Iraq's antiair capabilities and disrupt its communications.[6] Coalition aircraft flew more than 4,000 sorties from 1 to 20 March in this final surge before OIF began.[7]

Second, with a smaller number of ground forces available for this campaign than the Gulf War, the elements of speed and surprise had to offset lower troop levels.* By the time combat hostilities began, the United States had deployed 150,000 troops to support OIF, which was considerably lower than previously recommended troop levels.[8] By capitalizing on Coalition air superiority and executing elements of speed and surprise, 3d MAW planned for a two- to four-day separation between initial air and ground activities. Fixed-wing aircraft would aggressively attack enemy artillery sites, thus minimizing indirect fires as the 1st MarDiv advanced northwest to Baghdad. Believing a short period would suffice to shape the battlefield, ground forces could more quickly advance without jeopardizing the element of surprise.

Third, 3d MAW supported ground forces in its typical manner, while also operating in a nontraditional role as a separate maneuver element, protecting the eastern flank and diverting Iraqi forces away from the fight during the advance. Plotting a lengthy and challenging 400-mile route from staging areas in northern Kuwait to the heart of Baghdad, 3d MAW wrestled with logistical requirements to establish multiple remote air-ground operations sites. Further, the Marine Corps added another 100 miles to its journey when elements of 1st MarDiv moved north to Tikrit after the fall of Baghdad. As a result, 3d MAW flying squadrons and their support groups became more mobile than ever

*A sortie refers to one aircraft mission flown by a single plane.

*The small troop size to support OIF contradicted U.S. CentCom OPLAN 1003-98, which justified 380,000 troops for an Iraqi invasion.

Photo by LCpl Christopher H. Fitzgerald. Defense Imagery 030305-M-2237F-004

Marines assigned to VMFA(AW)-121 attach a ground line before refueling an F/A-18D Hornet on the flight line at Ahmed al-Jaber Air Base, Kuwait, before an Operation Southern Watch mission.

Photo by PO3 John Taucher, USN. Defense Imagery 030325-N-6610T-536

Marine Corps ordnance crews on board the amphibious assault ship USS Bataan *(LHD 5) prepare to upload GBU-12/B 500-pound Paveway II laser-guided bombs onto an AV-8B Harrier for an Operation Southern Watch mission.*

Photo courtesy of Col Charles J. Quilter II

A U.S. Naval Academy graduate and naval flight officer, Col Jonathan G. Miclot commanded VMFA(AW)-225 before his assignment as 3d MAW plans officer during Operation Iraqi Freedom.

imagined. Highlighting 3d MAW mobility, General Mattis declared that 1st MarDiv was one of the "most air-centric divisions" in the history of the Marine Corps.[9]

Fourth, a full arsenal of highly sophisticated precision-guided munitions coupled with a concept of rapid dominance through "shock and awe" was the overarching theme of the joint air strategy. In their book, *Shock and Awe: Achieving Rapid Dominance*, authors Harlan K. Ullman and James P. Wade introduced the concept of rapid dominance, which was designed to negatively affect the will, perception, and understanding of the adversary to respond by unleashing overwhelming and terrifying combined arms attacks.[10] Initially, joint aviation planners prepared for high altitude, targeted air strikes using joint direct attack munitions—one type of "smart" bomb. Called the "weapon of choice," precision-guided munitions enable pilots to fly at higher altitudes, thus minimizing the effect of antiaircraft missiles and artillery. Using precision-guided bombs represented a departure from Desert Storm, where the air campaign was dominated by "dumb" bombs—with no internal guidance mechanisms—which meant expending more ordnance to successfully neutralize targets. Significantly, the use of more ordnance did not equate to increased satisfactory bomb damage assessments.[11]

In the initial weeks of OIF, the Air Force flew a preponderance of high-altitude shock-and-awe type missions in the vicinity of Baghdad, while Marine aircraft flew mostly in southern Iraq, supporting 1st MarDiv and diverting enemy attention away from its eastern flank. Integrating a combination of firepower and precision-guided munitions aimed at destroying Iraqi equipment, command-and-control sites, and artillery targets created a deadly psychological dilemma for enemy forces—defend and be attacked or walk away from the fight.[12]

Planning, Planning, Planning

At the peak of combat operations, 3d MAW positioned 435 aircraft, plus 60 warplanes from the United Kingdom, at various Middle Eastern air bases and on board Navy ships. This impressive feat entailed an aggressive planning cycle that began in early 2002 in conjunction with 1st MarDiv. The efforts of air-ground team planners transpired at wing and division headquarters at Miramar and Camp Pendleton, California, and at two key Kuwaiti air bases—Ahmed al-Jaber and Ali al-Salem.[13]

As the senior Marine Corps aviation representative in CentCom's long-range planning element, 3d MAW plans director Colonel Jonathan G. Miclot frequently traveled to Tampa, Florida. The I MEF commander, Lieutenant General Michael Hagee, invited the lead planner in Korea, Lieutenant Colonel George W. Smith, to join Colonel Miclot so that the two officers could make a strong presence at the Marine Corps Forces Central Command headquarters facility. During initial planning sessions, the colonels realized the need to educate and promote the MAGTF model to an audience quite unfamiliar with one of its most appealing attributes—a flexible task organization coupled with building-block unit scalability. Once other services and conference attendees grasped the uniqueness of a ground unit with its own organic air support, future planning sessions moved to related discussions regarding where Marine aircraft, aircrews,

Photo by LCpl Andrew Williams. Defense Imagery 030819-M-UW798-005

The 3d MAW advance party established its forward command headquarters in Kuwait at Ahmed al-Jaber Air Base in November 2002 in anticipation of an official deployment soon after the holidays.

and wing staff would be based so they could provide support to the troops on the ground.[14] As a result of their success in highlighting the model's attributes, General Amos had no objections to embedding Marine liaison officers within joint agencies and commands.

Understanding that the operational, logistical, and staffing requirements to support OIF could not be adequately addressed without a physical assessment of the Gulf region, 3d MAW sent a small fact-finding team to Kuwait in late summer 2002. The assessment team began preparing for a massive forward deployment with an undetermined start date that ranged from a few months away to early 2003. The team gathered valuable data, surveying runways and touring bases, to evaluate existing infrastructures for future air operations. As new topics arose, the team added experts, such as administrators and liaison officers, who made follow-on trips to resolve issues and address manpower questions.

By November 2002, the air strategy was further refined when an advance party of 3d MAW commanders and senior planners departed California for Kuwait to complete final preparations for the movement of troops, aircraft, supplies, and equipment to the region. General Amos and a contingent of about 100 Marines celebrated Thanksgiving and Christmas in the Kuwaiti desert and awaited the official order to launch OIF. Those Marines who traveled with the advance party set up base camps and gained a firsthand view of how and where 3d MAW would deploy its assets.[15] The result was a master plan outlining the synchronized movement of an aviation combat element comprised of equipment, parts, and supplies to support 435 aircraft—along with 15,000 3d MAW personnel.

Assessing Current Capabilities: An Aging Aircraft Fleet

Marine Aviation at Headquarters Marine Corps managed a post–Gulf War transition plan that included the retirement of three legacy aircraft. Of those, two were jet-propelled: the Vietnam-era McDonnell Douglas RF-4B Phantom II photoreconnaissance plane and the Grumman A-6E Intruder attack plane. The functions of these aircraft were consolidated into a fighter-attack jet—the F/A-18 Hornet C and D models—which replaced them. The third aircraft retired was the turbopropeller-

Six Core Functions of Marine Aviation

Marine aviation functions have evolved over time, yet the core competencies have remained firmly rooted in official doctrine and at training commands. Air planners determined early in the process that each of the six doctrinal functions of aviation would be needed to support 1st MarDiv in OIF.

1. ***Offensive Air Support*** involves air operations conducted against enemy installations, facilities, and personnel to directly support the attainment of MAGTF objectives by destroying enemy resources or isolating enemy military forces. Its primary functions include providing fire and force protection through close air support and deep air support through air interdiction operations and armed reconnaissance or targets-of-opportunity missions.
2. ***Antiair Warfare*** involves offensive and defensive air actions to destroy or reduce the enemy air and missile threat to an acceptable level; to gain and maintain an appropriate degree of air superiority, as required; and to provide force protection.
3. ***Assault Support*** provides tactical mobility and logistical support capabilities to transport personnel, cargo, equipment, and supplies. It also includes the evacuation and tactical recovery of personnel and equipment, aerial refueling, and battlefield illumination within the area of operations.
4. ***Air Reconnaissance*** employs various aircraft platforms to support visual observation, multisensory collection, and electronic detection activities, primarily to acquire and assess intelligence information.
5. ***Electronic Warfare*** involves the use of electromagnetic and directed energy to attack the enemy and to provide protection and warfare support.
6. ***Control of Aircraft and Missiles*** integrates and manages the five other aviation functions, primarily to oversee centralized command and air direction and to maintain decentralized airspace control and management and the control of aircraft through the Marine air command and control system.[16]

driven North American Rockwell OV-10 Bronco, which was used primarily for tactical aerial observation and forward air control missions. Unmanned aerial vehicles (UAVs) and airborne forward air controllers (FACs) assigned to fixed-wing aircraft squadrons assumed the Bronco's functions.

By the end of the transition period in 2003, Marine aviation retained nine tactical aircraft types.* Although the fleet was aging, phased maintenance and planned avionics modifications, along with advanced weapons systems upgrades, maintained 3d MAW aircraft at the leading edge of performance and mission capabilities.

Fixed-wing aircraft that provided offensive air support and antiair warfare functions included the F/A-18C and D Hornet fighter-attack jet and the AV-8B Harrier II attack aircraft. The two-seat D-model Hornet proved invaluable when flying in advance of 1st MarDiv to support the aviation maneuver element, especially when conducting strike coordination and armed reconnaissance missions. In this role, a pair of D Hornets operated as a hunter-killer team, with one aircraft scouting for enemy targets as an airborne controller and the other striking and destroying them. Meanwhile, the Harrier II underwent significant changes that included engine upgrades and a new AN/AAQ-28(V) Litening II targeting pod, which allowed positive target identification and attack from altitudes of 20,000–25,000 feet as an alternative to low-level dive and pop-up targeting tactics.

For electronic warfare, 3d MAW loaned and shared its EA-6B Prowlers with the Air Force Coalition forces air component commander. Aerial reconnaissance was provided by combining attributes of fixed-wing aircraft and UAVs—primarily

*During OIF, 3d MAW managed tactical aircraft, excluding Marine Corps operational support aircraft (the McDonnell Douglas C-9 Skytrain, Beechcraft C-12 Huron, C-20 Gulfstream, and Cessna UC-35 Citation).

Photo by SSgt Matthew Hannen, USAF. Defense Imagery 030322-F-MY389-016

Although the Marine Corps' EA-6B Prowler was a sought-after joint aircraft because of its proven electronic warfare jamming and aerial reconnaissance capabilities, 3d MAW forwarded several mission requests to support the aviation combat element. The four-man aircrew assisted strike pilots by jamming enemy radar and communication sites, but they also collected tactical intelligence.

Hornets, Harriers with the Litening II pod, Prowlers, and AAI Corporation RQ-2B Pioneers (the Corps' UAV). Used for the first time during Desert Shield and Desert Storm, the RQ-2B Pioneer was flown remotely by an operator on the ground, thus eliminating possible aircraft crew losses.* The Pioneers played a role in influencing decision making through the skilled data interpretation and analyses of the digital images gathered on enemy activity.[17] Overall, because the Iraqi antiair threat remained relatively low, fixed-wing aircraft focused on armed aerial reconnaissance, deep air strikes, close air support, airborne forward air control, and air interdiction sorties, while KC-130 Hercules turbopropeller-driven planes conducted aerial refueling and cargo hauling missions and acted as an aerial control platform for the direct air support center.

Helicopters flew primarily offensive air support and assault support missions. For heavy lift cargo, troop transport, and medical evacuations, 3d MAW depended on its CH-53E Super Stallions and Boeing Vertol CH-46E Sea Knights. As attack and escort gunships, the Bell AH-1W Cobra* was armed with BGM-71 TOW—tube-launched, optically-tracked, wire command data link—and AGM-114 Hellfire guided missiles as well as with 2.75-inch Hydra 70 rockets and an M197 20mm cannon. The Bell UH-1N Iroquois (more commonly known as the "Huey" or "Twin Huey") utility and command helicopter could also be used as a gunship to carry rockets, a GAU-16 .50-caliber machine gun, and a GAU-17 7.62mm minigun or M240 7.62mm machine gun.

The sixth aviation function—control of aircraft and missiles—was satisfied through the Marine air command-and-control system and assets, resources, and personnel from a mix of Marine air control

*The Bell Boeing MV-22 Osprey tilt-rotor aircraft was still undergoing operational certification and was unavailable for use during OIF.

*The Bell AH-1W is officially named the Sea Cobra, which is virtually never used. It is almost universally known as the SuperCobra or Super Cobra, but will be referred to throughout this book, except when quoted, as the Cobra.

groups. By exercising centralized command and decentralized control, MACG-38 deployed as the lead group with a force of tactical air controllers, radar operators, and communicators, while its support partner—MWSG-37—secured and sustained permanent and forward air operations, ranging from Kuwait to north of Baghdad.

Complementing tactical aircraft weapons systems, night-vision capable 3d MAW aircraft allowed aircrews to use night-vision goggles. Although night operations were typically more complicated and dangerous, the goggles offered a greater tactical advantage from earlier wars. Additionally, precision guided weapons, such as joint direct attack munitions and laser-guided munitions, assisted pilots and aircrews when attacking and destroying targets. Crews were more likely to hit multiple targets accurately during a single bomb run, unlike in Operation Desert Storm where multiple aircraft flew strikes against single targets. Whether a target was a planned strike or one of opportunity, precision bombing increased the overall effectiveness and reliance on the Marine aviation combat element during OIF.

Making Critical Decisions

General Amos developed an air strategy that afforded Marine aviation an opportunity to plan and fight the air war as it had always dreamed—supporting Marines on the ground—yet also by operating as a separate aviation maneuver element.[18] Accordingly, the general and his assistant wing commander, Brigadier General Terry G. Robling, made a series of critical decisions during the planning process, and once in theater those decisions positively affected 3d MAW's execution of the air campaign. The rationale behind these decisions provides a greater appreciation of the inherent complexities and interrelated operational, logistical, safety, and manpower considerations associated with Marine aviation.

Aviation Operating as a Maneuver Element

Typically, the ground combat element of the MAGTF assumed the lead during training exercises and real-world operations, while the aviation combat element operated in a supporting role

A Marine Corps RQ-2A Pioneer unmanned aerial vehicle conducts a daytime reconnaissance mission over Blair Field during Operation Iraqi Freedom. The Pioneer provided ground commanders with real-time snapshots of enemy activity and the battlefield.

Photo by LCpl Andrew Williams. Defense Imagery 030823-M-UW798-015

for the ground commander's scheme of maneuver, providing lethal air firepower when needed. However, the idea of expanding this role and employing aviation as a separate maneuver element appealed to Generals Amos and Mattis.

Although not often used, the approach had been successfully tested and exercised in the Philippines during World War II. On 30 January 1945, U.S. Army General Douglas MacArthur ordered the 1st Calvary Division, commanded by Major General Verne D. Mudge, to race to Manila, free internees at the University of Santo Tomas, and seize Malacañan Palace. MAG-24 and MAG-32 provided a nine-aircraft detachment to guard the division's left flank. Patrolling from dawn to dusk overhead, Marine Corps aircrews searched roads and trails for enemy positions and activity. They then relayed critical intelligence information to General Mudge, recommending alternate routes that would bypass potential enemy engagements. Through this approach, Marine aviation provided invaluable reconnaissance data and air cover ahead of ground force movements while protecting the Army's flank.[19]

Sixty years later, reliance on Marine aviation as both support and an independent maneuver element generated a valuable force multiplier and allowed Generals Amos and Robling to execute a Corps air campaign. Attack helicopters provided assault and close air support to the division's immediate front, while tactical and reconnaissance fixed-wing aircraft flew farther out for deep targets, such as Iraqi artillery sites, surface-to-surface missiles, command posts, and armored tank columns. Doctrinally, close air support allows fixed- and rotary-wing aircraft to engage enemy targets that are near friendly forces. Its effectiveness requires detailed coordination and fires planning with the ground forces, whereas air interdiction is conducted farther away from friendly ground troops and aimed more at disrupting, delaying, or destroying the enemy's potential to execute decisive strikes.[20]

In essence, 3d MAW fixed-wing aircrews served as the eyes of the division, at times 60 miles in front of friendly troops, influencing command decision making and protecting Marines on the ground as they advanced toward Baghdad.[21] As a result, Marine aircraft "fixed" or held Iraqi military units in place. Just as had occurred in the Philippines, the 1st MarDiv traveled more quickly to Baghdad than anticipated with less disruption from the enemy forces that had been distracted from the main attack.[22]

Aircraft Disposition: On Land or at Sea

A lingering, yet basic, consideration for the wing's squadron commanders was where to base their aircraft. After months of negotiations, the Turkish government confirmed in February 2003 that neither the United States nor any Coalition forces would be allowed to use its air bases to launch offensive operations against Iraq. As a result, contingency planning for a northern border entrance into Iraq immediately halted. Without Turkish air facilities, Kuwait would be the primary base of operations for 3d MAW fixed- and rotary-wing aircraft. Unfortunately, in their current state, Kuwait's limited airfields simply could not accommodate the more than 400 U.S. aircraft and the hundreds more expected from Coalition nations. Further complicating the situation, planners suspected that minimal repairs had been performed on airfields inside Iraq since Desert Storm, so the wing faced the prospect of making major repairs to meet runway specifications along with expansion projects, such as parking ramps.

Given these constraints, wing planners devised an aircraft disposition plan that used a combination of Navy ships; existing air bases in Kuwait, Saudi Arabia, and Bahrain; and remote forward operating bases in Iraq. In Kuwait, two bases supported the bulk of permanent air operations—Ahmed al-Jaber for fixed-wing aircraft and Ali al-Salem for helicopters. Neither base, however, would be fully operational without major modifications. The 3d MAW survey team recognized that Ahmed al-Jaber needed significant expansion and runway upgrades to accommodate its aircraft. As a result, Navy Construction Battalions (Seabees) laid concrete ramps there to accommodate the air fleet. Once in theater, MWSG-37 joined the effort by building expeditionary airfields and a network of forward operating sites, all aimed at keeping pace with the momentum of 1st MarDiv. In spite of lacking much of their heavy equipment because of arrival delays, the Seabees provided essential runway construction resources and expertise that otherwise would have severely degraded MWSG-37 aviation ground support.[23]

Photo by LCpl Andrew Williams. Defense Imagery 030813-M-7837W-002

Aerial shot of a weapons storage area near Ahmed al-Jaber Air Base in Kuwait that was revamped for use during Operation Iraqi Freedom.

General Amos noted the magnitude of effort expended even before the first aircraft touched down in Kuwait. At al-Jaber, the wing had to build a parking ramp and a parking apron where five F/A-18 squadrons and one AV-8B Harrier squadron could be positioned. This requirement allowed the wing to accommodate 78 aircraft that would need to park and taxi and still be in compliance with safety regulations, since mechanics and technicians would need to load an assortment of ordnance payloads with thousand-pound bombs. Amos recalled, "The ramp was 1,300 feet by 950 feet, of poured concrete, one foot deep. When we ran out of money, we used AM-2 matting that is normally used for expeditionary fields and finally completed a 1,300-foot square parking area. We built new taxiways, buried fuel tanks, and placed fuel bladders out there, and ran fuel stations."[24]

The Hercules fleet was based at Bahrain with a small detachment in Kuwait, while Prowlers were colocated with Air Force squadrons at Prince Sultan Air Base in Saudi Arabia. The Harriers that initially launched from U.S. Navy ships in the Persian Gulf either returned to ship after completing missions or remained forward, operating deep inside Iraq from mobile forward arming and refueling points (FARPs). Aircraft assigned to the 15th MEU (SOC) remained with the USS *Tarawa's* (LHA 1) amphibious ready group in the Gulf waters off Kuwait's coast.

Direct Support Aircraft

To accommodate General Mattis's rapid advance, General Amos assigned helicopters and UAVs in direct support of the 1st MarDiv command post, each of the three regimental combat teams, and the 2d MEB. This allocated aircraft and aircrews to both avenues of approach to Baghdad, which enabled unit-focused intelligence gathering and dedicated air support. The air-ground team instilled a sense of camaraderie and afforded tactical commanders greater opportunity to request air support more efficiently for time sensitive missions, especially when presented with targets of opportunity.

Although unmanned aircraft were part of MACG-38, the unit's two squadrons split efforts and covered the division along its parallel routes. Marine Unmanned Aerial Vehicle Squadron 1 (VMU-1) primarily supported Regimental Combat

Team 1 (RCT-1) and Task Force Tarawa, while VMU-2 focused on RCT-5 and RCT-7.[25] The RQ-2B Pioneer flew a variety of reconnaissance missions, providing target surveillance of designated areas of interest, reconnaissance of helicopter approach and retirement lanes, and battle damage assessment.[26] The aircraft obtained detailed footage from its forward-facing infrared cameras, serving as the literal eyes of the ground forces.

The battlefield's distance inland from Kuwait prompted special consideration for aerial medical or casualty evacuations. Planners assigned two CH-46E transport helicopters to each regimental combat team to facilitate immediate evacuation missions. In a collaborative effort, 1st MarDiv began planning with the Navy's medical support staff, 3d MAW's lift support experts, and the 1st FSSG's administrative and logistical support branches to develop a swift, efficient process to extract casualties from as far forward on the battlefield as the tactical situation allowed. The Army also contributed to the direct support approach and offered dedicated access to a small fleet of Sikorsky UH-60 Black Hawk helicopters configured for medical transport.[27]

At least one Huey helicopter was located with General Mattis's command element and also with each regimental combat team as they advanced toward Baghdad. Crews flew primarily command-and-control flights during which ground commanders gained aerial situational awareness of the battlefield. Composite helicopter squadrons provided direct support to the I MEF combat missions, often separating into Cobra and Huey sections without a standard assignment to a specific regimental combat team. As an example, Marine Light Attack Helicopter Squadron 267 (HMLA-267) supported I MEF forces with surge and sustained operational sorties by forming five Cobra and five Huey combat sections or divisions. These task-organized combat crews provided 24-hour coverage to I MEF maneuver forces throughout southern and central Iraq.[28] One of the most important aspects of the direct support approach was that ground commanders had aviation assets readily available to meet immediate air support requests regardless of which group or squadron executed the mission.

Enhanced Tactics

The 3d MAW aviators and instructors not only tested new and enhanced weapons systems, but they also helped squadrons develop tactics that adapted to battlefield operations. The hunter-killer

The 1st MarDiv benefited from 3d MAW's approach of assigning aircraft in direct support of each of the three regimental combat teams. Here, a CH-46 Sea Knight waits near its supporting unit for a possible troop insertion, transportation, or immediate medical evacuation mission.

Photo courtesy of Col Charles J. Quilter II

Photo by Col Michael D. Visconage. Official U.S. Marine Corps photo

Aerial refueling, as demonstrated by this Marine KC-130 Hercules and F/A-18 Hornet, was a critical component of the air campaign that extended an aircraft's time on target.

tactics mentioned earlier employed by the crews of MAG-11's two-seat F/A-18Ds is one example. During combat operations, this technique reduced time-on-station for attack pilots and optimized limited aerial refueling time for the KC-130s. Aircrews worked directly with regimental combat team air officers to channel aircraft toward enemy armor in front of advancing ground troops, maximizing current imagery intelligence. This procedure led to several successful missions, most notably against the *Baghdad* and the *al-Nida Divisions*.[29]

Given Iraq's vast terrain laden with natural and manmade waterways, 3d MAW expected heavier reliance on helicopters during the march to Baghdad because rotary-wing aircraft could provide troop transport and aerial assaults of key bridges, intersections, and enemy defensive positions. Anticipating such missions, air planners altered tactics to accommodate an anticipated surge in assault support requests. MAG-16 and MAG-39, along with 1st MarDiv, collectively designed a battalion-sized reusable helicopter lift plan tailored for specific scenarios. The lift package included a set of predefined launch criteria, including minimum weather conditions, that aided commander's tactical planning and decision making.

Additionally, urban close air support was a high-interest topic and the wing spearheaded several expeditionary force-level command conferences and planning sessions to discuss how to best employ aircraft in such environments. They hoped to prevent situations where low-flying helicopters could become easy targets as happened in Mogadishu, Somalia, in 1993. One of the lead proponents of this initiative was Colonel Paul K. Hopper, the wing air officer, who hosted an urban close air support symposium from 6 to 10 January 2003. Focusing on urban fires, Colonel Hopper invited experts from information operations, targeting, collections, Marine Aviation Weapons and Tactics Squadron 1 (MAWTS-1), the Marine Corps Warfighting Laboratory, and the Rand Corporation. This group developed a framework that was later implemented to seize Baghdad.[30] Believing that the capital would be heavily defended, 3d MAW expected significant enemy opposition. In anticipation of that situation, low-flying attack helicopter aircrews, using precision ordnance,

Photo by SSgt Sean M. Worrell, USAF. Defense Imagery 001030-F-FD742-007

An aerial view of the combined air operations center at Prince Sultan Air Base, Saudi Arabia, where joint service tactical air controllers enforced the Iraqi no-fly zone during Operations Southern Watch and Iraqi Freedom.

needed refined tactics for use over a city with five million people.

Understandably, the concern centered on coordinating fire so that crews could provide coverage for Marines on the ground yet not place their aircraft in vulnerable positions where the enemy would have an easy target. The symposium addressed likely scenarios in which helicopter crews would engage the enemy from a few hundred feet above rooftops only a few stories high. This scenario was risky for aircrews and ground troops alike.[31]

Cadre of Senior Liaison Officers and Battle Captains

General Amos sought to foster sound relationships by handpicking a team of the most qualified staff officers to serve throughout joint service commands, Air Force air control agencies, and I MEF staff, as well as in support of regimental combat team commanders. These liaison officers were aviation experts who communicated directly with battlefield commanders and joint cell decision makers. By providing aviation-specific technical information and cultivating close working relationships during the planning and deployment phases, bonds formed, promoting a better understanding of Marine air-ground operations. Although assigning liaison officers was not a new concept, the general's strategic placement and extensive use of these officers provided him with direct access to information about joint air operations that made their employment singularly effective.

One newly created liaison billet of particular interest to General Amos was Colonel Martin Post's assignment in the combined air operations center at Prince Sultan Air Base. As the senior Marine, Colonel Post's primary role was to act as the general's on-site advisor for all service air matters, and his second duty entailed oversight and mission assignment of all 3d MAW aircraft listed on the joint air tasking order. By having a well-versed naval aviator readily accessible, General Moseley's staff could make more informed decisions about Marine-specific aircraft allocation. Although the Air Force controlled the strategic air campaign, Marines also engaged in fighting the bigger war by flying excess sorties as directed by joint air command agencies. Once combat operations began, it was not uncommon for Air Force Fairchild Republic A-10 Thunderbolt II attack planes* to support Marines on the ground, particularly during intense firefights such as in the battle of an-Nasiriyah, while Hornets flew deep strike missions into Baghdad. In the end, OIF demonstrated that the Corps could retain control of its aircraft, yet successfully comply with Air Force protocols, and operate in a melting pot of shared aviation resources, joint targeting schedules, and coordinated command-and-control requirements.[32]

In January 2003, the 4th MAW commander, Brigadier General Harold J. Fruchtnicht, mobilized reservists from his command to form a liaison cell that deployed to the joint forces air command center at Prince Sultan Air Base and filled senior billets at the director and deputy director levels.[33] MACG-38 also recruited for an exchange of liaison officers from other command-and-control agencies and assigned them to the Air Force combined air operations center, the 363d Expeditionary Airborne Air Control Squadron's airborne early warning and control system, air support operations center, and Kuwaiti air traffic control agencies. The efforts of these liaison officers ensured full integration of the Marine style of centralized command and decentralized control throughout the campaign, especially during heated combat operations.

Another agency in which General Amos modified the responsibilities of a liaison was in the creation of a new billet within the MACG-38 tactical air command center—the "battle captain." Relying on the tactical air command center as his primary command post to view current operations and envision future operations, General Amos studied its basic functions before the air war began and sought improvements. Seeking the advice of a handful of retired Marine mentors who cautioned about becoming too bogged down in the current fight, the general realized that he also needed to focus several days ahead on the future air war.[34] Although his senior watch officer billets were staffed with trained tactical air controllers, he understood that they could easily become saturated with too much data, which in turn could impede the time-sensitive flow of information to commanders and staff officers. Anticipating that 3d MAW would use helicopters extensively for tactical missions—primarily for close air support—General Amos wanted to safeguard against his senior watch officers inside the tactical air command center from becoming overburdened with strategic matters while trying to forward critical flight information to the pilots, FACs, ground units, and internal controllers. As an alternative, Amos chose four experienced aviators to fill the battle captain billets as a means of distributing the workload. This placed four senior aviators in charge—instead of controllers—as the senior watch officers. Battle captains held the rank of colonel and worked 12-hour rotating shifts that allowed them to share the responsibility of managing the joint airspace to support current operations while also planning for future air operations. Battle captains held air controller mission briefings, selected targets, assigned missions, and maintained control of air and surface fires, whereas the senior air director and watch officers concentrated on specific missions that entailed directing aircraft, separating aircraft at defined altitudes, and monitoring the airspace.[35]

General Amos continually reemphasized his commitment within the air wing to support 1st MarDiv, sharing this sentiment during a conversation with the division commander: "I told General Mattis that there would be literally nothing, within reason, that I would not do to take care of his Marines on the ground. Once the planners and liaison officers figured out the two commanding generals were always going to agree with one another, the staffs got it. They understood that the relationship between the two commanders was very important."[36]

Forward Air Controllers

Employing Marine FACs—pilots assigned to ground units who control tactical aircraft—during OIF demonstrated one of the primary advantages of the Marine Corps air-ground team that made it unique from the other services. General Amos selected 16 experienced FACs to augment each I MEF ground maneuver unit to offer immediate aviation expertise to the ground combat element. Air planners searched for a core group of majors and lieutenant colonels with credentials that included combat experience and a previous FAC tour. Those Marine pilots who served as aviation liaison officers supplemented unit commanders with insight about

*The A-10, usually referred to by its nickname the "Warthog," is a rugged, well-armed, and heavily armored aircraft designed exclusively for attacking vehicles and troops on the ground. It is famous, or in some eyes, infamous, for its role in close air support given previous involvement in friendly fire incidents.

The Forward Air Controller

The FAC's role has been engrained into MAGTF operations for decades and, over the years, modeled by other services. The FAC's primary responsibilities are as follows:

- know the enemy situation, selected targets, and location of friendly units;
- know the supported unit's plans, position, and needs;
- validate targets of opportunity;
- advise the commander on the proper employment of air assets;
- submit immediate requests for close air support;
- control close air support with the supported commander's approval; and
- perform battlefield damage assessment.[37]

Collectively, these measures were intended to prevent friendly fire incidents by ensuring accurate bombing of designated enemy targets. As more sophisticated communication systems along with tactics, techniques, and procedures became available, the Department of Defense released *Joint Pub 3-09.3: Joint Tactics, Techniques, and Procedures for Close Air Support (CAS)* in 1995—with a major revision in 2003—to provide uniform procedures for close air support.*

*Headquarters closely monitors FAC eligibility and training requirements, which include having at least two years operational flying experience and being a graduate of the expeditionary warfare training group and the tactical group and the tactical air control party course. After training completion, graduates earn a FAC military occupational specialty (MOS 7502) and are considered certified joint terminal attack controllers.

how to best employ aviation assets, ordnance, and tactics while integrating aircraft capabilities into the ground scheme of maneuver. Their role was to maintain continual situational awareness; know the location of "friendly" units; and understand the rules of engagement to prevent attacks on noncombatant civilians and off-limit buildings, such as religious sites or hospitals.

Force Augmentation and Mobilizing Reservists

Aviation planners realized early in the process that they needed more pilots, liaison officers, maintenance technicians, aviation logisticians, tactical air controllers, and ground support personnel to adequately execute 3d MAW's dual role during OIF.

Readily identifying shortages, the wing asked manpower experts at Headquarters Marine Corps to search for additional units, squadrons, and individuals to supplement its rosters. Although finding resources for squadron deployments was a common practice at the headquarters level, staffing for such a sizeable move strained the routine planning process.* On 14 January, Headquarters Marine Corps authorized a stop-loss action under Marine Administrative Message 007/03 that temporarily stabilized the workforce so that manpower planners could match operational requirements with personnel resources. The action halted normal personnel turnover and prevented Marines from leaving active duty, retiring, or transferring to other billets without approval. This decision allowed manpower planners to prioritize unit and individual personnel requirements for a specific military occupational specialty and aggressively seek reservists with specialized civilian skills.[38]

Marine Corps Forces Reserve in New Orleans, Louisiana, aided 3d MAW planners and began activating squadrons, aviation attachments, and individual reservists with aviation occupational and civilian skills. The mobilization process varied as reservists left government and civilian employers to join their respective commands, often with only a few days notification before deployment. Reserve Marines from across the country reported to active duty under Deployment Order 177B, which restricted their employment to Southern California to fill critical billets vacated by deploying active duty Marines. Movement of Reserve Marines outside the continental United States did not occur until the secretary of defense authorized a sepa-

*A deployment of this magnitude, nearing 15,000 air wing personnel, had not occurred since the Gulf War in 1990–91.

Photo courtesy of BGen Robert E. Milstead Jr.

Gens Amos and Robling meet with 3d MAW commanders during a planning session. Pictured from left to right are Col Ronnell R. McFarland, Col Michael C. Anderson, Col Richard W. Spencer, BGen Terry G. Robling, MajGen James F. Amos, Col Mark R. Savarese, Col Randolph D. Alles, Col Robert E. Milstead Jr., and Col Stuart L. Knoll.

rate deployment order. In the end, 3d MAW reinforced 30 percent of its total troop strength from the reserve community.

General Amos's augmentation requests were ultimately met with additional personnel, aircraft, and equipment from two other wings: 2d MAW in Cherry Point, North Carolina, and 4th MAW, a reserve air wing in New Orleans, Louisiana. This augmentation added MAG-29 from New River, North Carolina, to the air-ground team, which provided a much-desired composite helicopter group to enhance 3d MAW's mix of tactical, utility, and transport rotary-wing assets. Now reinforced with seven groups, 3d MAW included MAGs 11, 13, 16, 29, and 39, coupled with MACG-38 and MWSG-37. Adding to the short-range antiair defense and force protection capability, 2d and 3d Low Altitude Air Defense Batteries provided convoy and ground security at forward operating bases.

Additionally, joint service augmentation from the U.S. Army strengthened the air wing in terms of medical evacuation by supplying 15 UH-60 Black Hawk helicopters from the 498th Medical Company (Air Ambulance) and its air defense and security support with elements of the 108th Air Defense Artillery Brigade providing Patriot missile security.

Even though 3d MAW successfully retained operational control of its organic aviation assets during this campaign—a major victory in itself—a myriad of related complex challenges arose as the air strategy became more concrete. Some of these centered on oversight of joint airspace operations, pilot rules of engagement, and aircraft performance and readiness, while others were less controllable, such as climate and weather. Regardless, wing planners sought innovative ways to mitigate such challenges and execute the air campaign as intended.

Photo courtesy of Col Charles J. Quilter II

A graduate of Central Washington University, BGen Terry G. Robling flew the F-4 Phantom before his selection as one of 24 officers to the F/A-18D Hornet transition board. Throughout his career, he held various operational and staff positions and served as 3d MAW's assistant wing commander during Operation Iraqi Freedom. Gen Robling remained in Iraq until all of his Marines and aircraft returned home to California in November 2003.

Joint Air Tasking in the "Purple Haze"

The hallmark agreement between Generals Moseley and Amos during the planning phase clarified the command-and-control relationship between the Air Force and the Marines and became the heart of the Corps' aviation strategy throughout its execution. Recalling the more bureaucratic approach used during Desert Storm in which the Air Force approved all Marine air requests, attempted to control all aviation assets, and assigned aircraft before aircrews could acquire targets, this campaign afforded an opportunity to resolve lingering issues from the past.* As a result, a genuine joint airspace relationship evolved—referred to as the "purple haze"—signifying the epitome of interservice collaboration. Without question, General Moseley still owned all assets in the joint area of operations, including those of the Marines, yet General Amos retained operational and tactical control of 3d MAW aircraft to support Marines on the ground. Thus, formal operational control of Marine air assets never transferred from Marine Corps to joint control.[39] In simple terms, the Marine Corps leased its airspace from the Air Force.

General Robling recalled one particular meeting—prior to 3d MAW's deployment forward—in which General Moseley clarified any confusion about who would control the joint airspace and the daily flight schedule or air tasking order. Amidst the chatter in a room occupied with senior-level service members, General Moseley stated firmly, "I'm the joint forces air commander. I own from one half inch above the ground to infinity of the area of operations and I'm not giving that up. But what I will allow is the predominant, the Marine Corps to put their air over the Marines just like they always do and we'll work out the excess sorties."[40]

Determined to enforce this agreement during the air campaign, General Amos depended on the Marine air liaison officers within General Moseley's air components, specifically in the combined force air command center, to serve in advisory roles. Comparable to the Marine Corps' direct air support center, the Air Force's command hub employed a joint staff of approximately 2,000 personnel by mid-March.[41] One of the most difficult hurdles to overcome was assigning planned missions to multiservice aircraft and listing them on the joint air tasking order. Similar to a master flight schedule, the air tasking order was published daily and listed preassigned sorties with corresponding mission information, such as specific pilot-controller instructions, call signs, targets, controlling agencies, and estimated launch times.[42] At the peak of air operations, the tasking order was distributed to controlling agencies and subordinate units based on a projected, strategic 96-hour planning cycle. It often exceeded 100 pages.

The problem for the Marine Corps was anticipating a list of potential on-call missions based on ever-changing operational tempos, rather than projecting preplanned missions four days in advance. Just like the Moseley-Amos agreement,

*Although a layered, formal approval chain worked well for strategic-level air campaigns like those flown by the Air Force, the Gulf War approach jeopardized immediate tactical decision making among Marine air controllers, pilots, and commanders, leaving minimal flexibility. Such a cumbersome approval process would bog down response times and hamper on-call missions to support MajGen Mattis's rapid advance to Baghdad.

joint air planners reached a compromise that allowed Marine missions to be listed or accounted for twice—first on the master or theater-level air tasking order and second on a Marine-specific direct support air tasking order. This allowed Marine aviation to retain its flexibility to support on-call missions as a maneuver element, yet also comply within the 96-hour requirement, thus minimizing a labor-intensive reconciliation effort at the end of a shift comparing preassigned missions with detailed usage data of actual sorties flown. To further simplify the assignment workload, Marine helicopter missions were listed on the air tasking order as high-level placeholders without mission details, which allowed further flexibility. In turn, the direct air support center could more closely direct and oversee helicopter missions because pilots would often have numerous take-offs and landings during a single mission. Although this approach was far from perfect and never totally resolved, it eased the administrative burden of tasking and tracking joint assets and minimized a massive postmission reconciliation effort.

To generate the daily air tasking order more efficiently, the Air Force developed a relatively new automated tool—the theater battle management core system. Navy and Marine Corps liaison officers working in joint air control agencies were introduced to the system for the first time in March 2003, which resulted in an initial steep learning curve as users became familiar with software functionality and capabilities. Marine Tactical Air Command Squadron 38 (MTACS-38) supported 3d MAW and I MEF with air tasking order production by writing, disseminating, and executing it on a daily basis. This included a network in excess of 90 desktop computers coupled with 43 remote and 72 web-based remote terminals.[43]

Understanding the Law of War and the Rules of Engagement

When loaded with precision-guided munitions, the fleet of legacy, yet sophisticated, aircraft not only ensured target destruction but had the secondary psychological effect of disrupting enemy resistance and destroying their will to fight. But neither role could occur without compliance with the rules of engagement.

Although the burden of publishing clear and concise rules of engagement—self-imposed limitations governing the use of force—rested with CentCom, 3d MAW planners, pilots, and controllers had to conduct the air campaign within the confines of the law of war—the international law that binds nations by treaty or by custom. Any psychological advantage or goodwill gained in the air campaign could be lost if aircraft attacked noncombatant civilians or sites.

Although speed and surprise were critical for OIF's success, Marine aviation had to minimize the potential for incidents involving friendly fire or killing noncombatants. The rules required that Coalition forces positively identify a target as enemy before engaging it to preserve the country's infrastructure and reduce the need for reconstruction after the war's conclusion. As the war evolved, aviators understandably found many of the constraints frustrating and challenging, especially in the face of an enemy who, contrary to the laws of war, used medical and religious sites for military purposes and wore civilian clothing to blend in with noncombatant civilian populations. Nonetheless, 3d MAW aircrews maintained the delicate balance to maximize lethality while minimizing collateral damage. As a means of further understanding the laws of war and the rules of engagement, 3d MAW reinforced their importance during training sessions.

Marines from the wing's advance party deployed in small groups to Kuwait to cultivate friendships with neighboring Coalition-friendly countries. One of the team's first assignments was to lay out the functional areas of the tactical air command center where battle staff training would occur a few months later. Training sessions primarily centered on rules of engagement, but public affairs and media handling techniques; first aid; and procedures for joint tactical air, assault support, and medical evacuation requests were covered as well. Once in theater, 3d MAW participated in two CentCom exercises to prepare for joint operations: Lucky Warrior 3-1, which focused on the combined forces land component commander perspective, and Internal Look, which focused on wargame exercises.

Adapting to the Climate and Maintaining Aircraft Performance

Iraq is known for its frigid winters and hot, dry summers that often exceed 100 degrees Fahr-

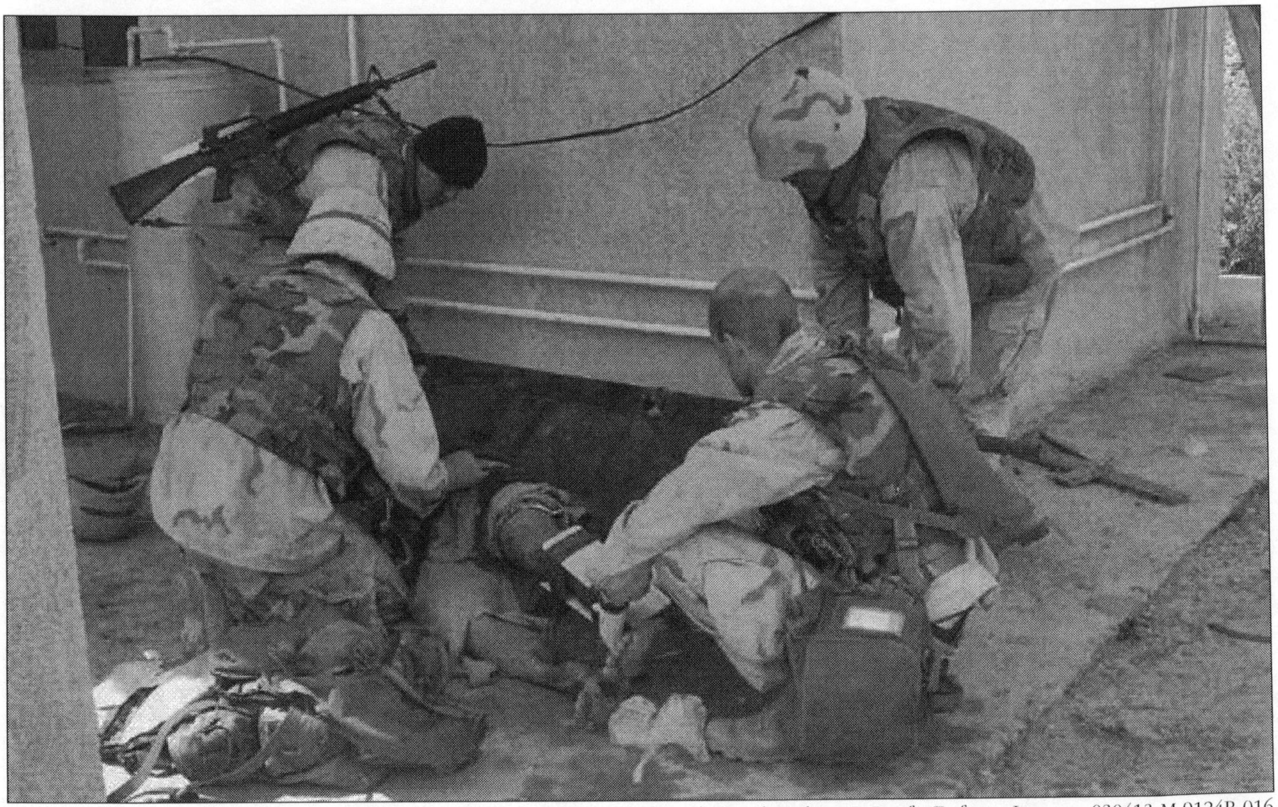

Photo by LCpl Andrew P. Roufs. Defense Imagery 030413-M-9124R-016

Navy hospital corpsmen assigned to 1st LAR, 1st MarDiv, attend to a wounded enemy combatant who wore no uniform. This disregard for international convention made it more difficult to authorize enemy targets during the fighting in Iraq.

enheit from May to September. Although not as severe as the climate experienced by those who trained at Yuma, Arizona, and Twentynine Palms, California, the brutal summer months in Iraq are defined by two wind patterns: the *Sharqi* and the *Shamal*. From April to November, dusty Sharqi winds with occasional gusts up to 80 kilometers per hour that can carry sands several thousand feet high dominate the southern part of the country. The Shamal phenomenon produces a steady prevailing wind from mid-June to mid-September that brings blinding sandstorms followed by another Sharqi in the fall.[44]

Under such extreme conditions, flight operations and aircraft performance would be adversely affected, so 3d MAW Marines had to adapt. Pilots learn early in their training that high air temperatures cause aircraft to have a poor rate of climb and demand more runway length for takeoff and landing. In general, aircraft perform best in colder temperatures. Seasoned maintenance personnel anticipated the negative effects that Iraq's extreme heat and harsh desert environment would have on aircraft readiness.* Of greatest concern were the violent winds and blowing dust storms, which limited pilot visibility. Sand and foreign object debris created a ripe environment to reduce aircraft performance and increase maintenance cycles. Keeping grit out of jet engines and deep inside airframe crevices was nearly impossible. Coupled with an aging fleet, Iraq's environment was not conducive to high-performance machinery despite regularly scheduled maintenance.

Although climate and weather could not be controlled, 3d MAW's aviation ground support, logisticians, and maintenance personnel worked together to mitigate as many issues as possible. The weather section from MWSG-37 scheduled two daily forecasts instead of one for General Amos and the battle staff that coincided with intelligence briefings. Complementary information not only kept commanders abreast of enemy threats and changing weather conditions, it provided air-

*Overall, airfoils—wings and rotor blades—perform better (i.e., generate lift) when air density is greater as with colder temperatures and at lower altitudes.

Photo courtesy of Col Charles J. Quilter II

Dust storms and fine sand played havoc with 3d MAW aircraft and maintenance crews. This dust cloud temporarily dominated the primary Marine fixed-wing air base at Ahmed al-Jaber, Kuwait.

crews with the most current situational information before mission launch. From a maintenance perspective, lessons learned during the Gulf War now affected how squadron and intermediate maintenance Marines prepared for operations in an unforgiving desert environment. Two relatively simple and inexpensive preventive maintenance practices—frequently washing airframes and applying lubricants to certain components—paid great dividends. With adverse weather in mind, 3d MAW wanted three good flying days before the ground war started to finish shaping the battlefield and for assault support missions. Unfortunately, even on the first night of the air campaign, the weather never cooperated.

Blending Marine Aviation and Logistics Squadrons

With more than 400 aircraft of 9 different types and multiple models in a harsh desert climate needing maintenance and logistical support, Colonel Gilbert B. Diaz, 3d MAW deputy chief of staff for aviation logistics, required a simple yet flexible plan. He and his staff published a plan outlining the aviation supply concept and logistical support requirements to maintain the highest possible aircraft readiness rates, both full- and partially mission capable. Part of the plan was to establish an intermediate maintenance capability on board two U.S. Navy aviation logistics support ships, the SS *Wright* (T-AVB-3) for fixed-wing aircraft and the SS *Curtiss* (T-AVB-4) for support helicopters. (T-AVB is the designation for a roll-on/roll-off container ship conversion dedicated to moving a Marine aviation logistics squadron by sea.*) Typically, each Marine

*A 20-year-old idea, the Marine Corps authorized a feasibility study for an aviation logistics support ship in November 1985, and the Navy converted two cargo container ships to fulfill this mission. During this time, the Marine Corps aviation logistics community introduced the Marine aviation logistics support program, which defined various contingency support packages of aircraft parts based on several factors, including past usage data and the length of deployment. The program outlined inventory levels of aircraft materials, including repairable components and consumable parts, that would complement aviation combat element requirements and the flexibility of MAGTF operations.

Photo courtesy of Col Charles J. Quilter II

Col Gilbert B. Diaz, 3d MAW's deputy chief of staff, Aviation Logistics Department, deployed with Marine aviation logistics squadrons. A seasoned logistician, he recognized that the ultimate testimony of aircraft readiness was performed behind the scenes by Marines whose sole objective was "fixing and flying" the wing's aircraft.

aircraft group, through its Marine aviation logistics squadron, provided intermediate-level maintenance functions—a prerequisite maintenance phase before full depot-level repair—along with logistical and ordnance support to its squadrons.* Such functions included work centers to repair engines, airframes, and avionics and to perform quality assurance tasks.[45]

Anticipating a deployment order soon after the New Year in 2003, Marine aviation and logistics squadrons prepared for potential aircraft readiness issues if the timeline extended major air operations into late spring and early summer. Colonel Diaz's foresight encompassed building a robust parts inventory, placing engine pools at the main air bases, and establishing intermediate repair facilities on board Navy ships. Based on Diaz's 30 years of experience and guidance, 3d MAW deployed forward with 100 percent accountability, or inventory validity, of 6,219 fixed- and rotary-wing repairable and consumable parts categories, which was a difficult metric to achieve given the aircraft mix.[46]

Although six detachments of Marine aviation logistics squadrons deployed to support OIF, Colonel Diaz blended East and West Coast units, each with unique asset management philosophies and business processes, into a streamlined operation. Further, he consolidated assets, resources, and personnel to form two dominant units: Marine Aviation Logistics Squadron 11 (MALS-11) for fixed-wing aircraft and MALS-39 for rotary-wing aircraft. Once the war began, al-Jaber and Ali al-Salem Air Bases were the hubs while the *Wright* and *Curtiss* were the major spokes of maintenance and supply activities coupled with forward operating sites.

The inherent complexity of aviation logistics materials management and its well-defined supply chain demanded detailed coordination and vendor agreements with commercial transportation carriers that would deliver critical avionics components and high-demand replacement parts. In turn, expediting parts required aggressive procurement oversight from military and commercial stock points, multiple modes of transportation, and interagency liaisons to expeditiously replenish, track, and ship parts from the United States to Kuwait. By establishing commercial air delivery contracts well in advance of the forward deployment, aviation logistics squadrons used overnight delivery services to fly parts directly to al-Jaber and Ali al-Salem. This approach contributed to maintaining acceptable fixed- and rotary-wing aircraft readiness rates during the march to Baghdad, despite austere forward bases and harsh weather conditions.

Colonel Diaz's staff also oversaw daily Marine logistics flights to and from the two maintenance ships and all bases in Kuwait, Saudi Arabia, and Bahrain with a fleet of heavy-lift helicopters and an assortment of available cargo aircraft.[47] Once in theater, however, one of the colonel's most frustrating problems was moving parts from one location to another because of scarce ground transportation. Although the situation improved over time with reliance on KC-130s to transport

*In 1985, a major organizational change spearheaded by Col Donald E. Davis revised the command structure for aviation logistics and maintenance Marines, changing it from Headquarters and Maintenance Squadrons to Marine Aviation Logistics Squadrons (MALS). The new structure offered a defined career path for aviation logistics and maintenance officers, allowing them to serve in squadron commanding officer and executive officer billets.

Photo by Sgt Nicholas S. Hizer. Defense Imagery 030211-M-5266H-007

Cpl Kristen Myrick with VMFA(AW)-225 checks a safety pin on an AIM-9 Sidewinder missile attached to an F/A-18 Hornet at Ahmed al-Jaber Air Base, Kuwait, in February 2003.

parts, Colonel Diaz summarized it best: "It's frustrating when you have a $100,000 part, and you can't move it to a site to fix an aircraft."[48]

When 3d MAW began flying 250–300 sorties a day, squadron maintenance crews understood the need to inspect, repair, and return aircraft to the flight line quickly while also maintaining quality and safety standards. Although the entire air wing fought a never-ending battle with sand and grit, helicopters and the propeller-driven C-130s presented even more issues because of blade erosion. The squadrons used blade tape, but the problem was never fully resolved. Chief Warrant Officer-2 Kraig A. Meyer, HMLA-267 maintenance control chief, repeatedly dealt with desert sand on the squadron helicopters while trying to perform phased maintenance under time constraints to return broken aircraft to flight status as quickly as possible. For UH-1N helicopters, an inspection took 6 hours, while an engine wash was slated at 25-hour intervals. Both efforts pushed Marines well beyond eight-hour work shifts to keep the aircraft flying. The maintenance chief prioritized the workload following simple guidelines, "It was basically, fix the communications systems first, fix the weapons systems next, and shoot 'em back out the door."[49]

Chapter 4
Controlling and Supporting Aircraft

Managing the Airspace with Marine Air Control Group 38

Dual role planning for the 3d MAW demanded that MACG-38 maintain a tactical communications network over greater distances than ever before while retaining command and control of the I MEF airspace. Most of this daunting task fell upon Colonel Ronnell R. McFarland, commanding officer of MACG-38.

Colonel McFarland relied on his 30-year tenure within Marine air command and control to recognize that his group could not effectively accomplish such enormous tasks without support. What emerged was a natural partnership with MWSG-37 as the two units planned and established forward operating sites during 1st MarDiv's advance to Baghdad. Through their efforts, MACG-38 and MWSG-37 consolidated assets, personnel, and talents, becoming the foundation of the wing's mobility that ultimately stretched more than 450 miles from the northern border of Kuwait to Tikrit.

A few months prior to deploying forward, Major Jeffrey P. Davis, MACG-38 assistant operations officer, focused on the sustainment effort of the control group, ensuring that Marines completed weapons qualifications and received training on tactical convoy security, rules of engagement, and map orientation. Once the air campaign began, he experienced forward mobility firsthand while overseeing a critical component of extending forward tactical air communications. He led the reception, onward movement, staging, and integration of the group's men and materiel at Jalibah and an-Numaniyah in Iraq. Coordinating with MWSG-37, Davis was instrumental in merging a combined forward command element in which the air wing's two nonflying groups could plot the aviation ground scheme of maneuver. The arrangement simply made sense and facilitated connectivity at FARPs and at forward operating bases (FOBs). Although the original plan included 10 to 12 forward sites, 21 FARPs were eventually created to keep up with the fast-paced operational tempo and evolving needs of the ground combat element.

Major Davis believed his experiences as a veteran of Operations Desert Storm and Enduring Freedom and as an instructor at Marine Aviation Weapons and Tactics Squadron 1 in Yuma, Arizona, had groomed him for the challenges of readying control group squadrons for deployment and sustaining forward air operations. He recalled, "It was more than just nuts and bolts to get lance corporals ready for deployment. In the end, during hostilities, I was absolutely awed by [what] Marines can do when needed; the teamwork throughout the air wing was just incredible."[1]

MACG-38 received personnel reinforcements and equipment augmentation from two other groups: MACG-28 of the 2d MAW and MACG-48, a reserve unit from 4th MAW. Once in theater, MACG-38 assumed initial operational control of

Notional Marine Air Control Group Organization

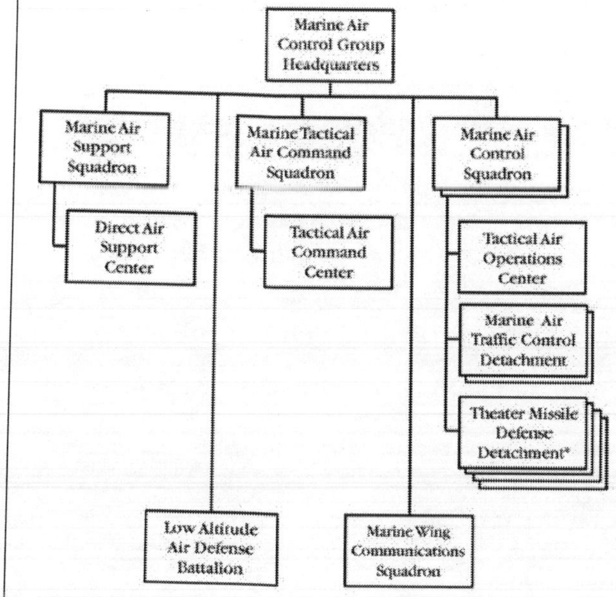

*There are ony four theater missle defense detachments in the Marine Corps active duty structure.

the Army's 108th Air Defense Artillery Brigade that employed the Patriot missile. Its soldiers were assigned to al-Jaber Air Base in Kuwait with a liaison officer positioned inside the tactical air command center. Just before the start of combat operations, I MEF assumed direct operational control of the brigade. The missile unit's mission benefited from full integration into the Marine Corps command-and-control system by receiving critical early warning cues from the tactical air operations center that helped shape launch decisions.[2] Once fully staffed, MACG-38 was poised to perform its primary mission: to establish and sustain the Marine air command-and-control system on behalf of 3d MAW.[3]

Extending Tactical Communications

Anticipating an official deployment order, McFarland sent 110 air control group Marines to Kuwait with the 3d MAW site survey team to explore options for establishing both fixed and mobile tactical communications. The team's assessment paid huge dividends as MACG-38 orchestrated a master plan to meet operational and logistical needs while complying with Marine air control doctrine—to exercise centralized command through the tactical air command center and decentralized control primarily through two control agencies, the direct air support and tactical air operations centers.* The colonel used existing equipment configurations and upgraded software applications, such as installing a military version of Internet chat or instant messaging software, that enabled real-time communications between aircrews and controllers during missions.

Using the new generation of Internet-enabled digital and satellite communications meant that MACG-38 bottom-line priorities were twofold: maintain connectivity at al-Jaber and Ali al-Salem in Kuwait, and establish communications at forward air sites inside Iraq once the war started. As a result, they had one of the most mobile and sophisticated digital-based air command-and-control systems implemented in the history of the Marine Corps that stretched from Bahrain, Saudi Arabia, Kuwait, and throughout Iraq. Marine Wing Communications Squadron 38 (MWCS-38) designed a "nodal communications" architecture that included satellite capability and terrestrial microwave communications equipment to geographically link outlying sites.*

At the Tactical Air Command Center (TACC) at al-Jaber, Kuwait, communicators relied on a digital technical control system and a tactical data network gateway coupled with AN/TTC-42 automatic telephone central office switching with miles of cable.[4] To maintain oversight at remote sites, the squadron depended on its secure, deployable command element "93 Jump Package." Its capability was absolutely vital to the employment of Marine aviation as a maneuver element because forward communications provided real-time reconnaissance data and images from the UAVs directly to the tactical air command center where General Amos's battle staff directed current and future air operations.

Decentralized Control

Reliability of the extended tactical aviation communications network was validated daily by its users—ground commanders, pilots, forward air controllers, and tactical air controllers. Before hostilities began, Colonel McFarland made two pivotal decisions that further demonstrated 3d MAW's mobility, decentralized control, and responsiveness to the ground scheme of maneuver. These decisions cost nothing to implement and necessitated only a reallocation of resources, roles, and responsibilities for two MACG-38 air control agencies—the direct air support and tactical air operations centers.

First, instead of staffing one centralized direct air support center to monitor all air control traffic originating from Kuwait to Baghdad, as was typically done, the colonel formed three unit-oriented air support elements and assigned them to each of 1st MarDiv's maneuver elements.** The main direct air support center was outfitted with an AN/TSQ-207 communications air support central platform,

*The interdependent agencies within the Marine air command-and-control system provided the Marine aviation combat element commander with the capability to monitor, supervise, and influence the application of the six functions of Marine aviation.

*As a means of distributing the initial workload, Marine Air Support Squadron 3 (MASS-3) and MTACS-38 jointly established and maintained the Tactical Air Command Center (TACC) and the direct air support center, while the Marine wing communications squadron installed, operated, and maintained the initial tactical communications nodes at 3d MAW's permanent air bases in Kuwait.

**Typically, the TACC included one close battle cell whose staff monitored the current fight and communicated with one direct air support center; however, given the scope of operations, the joint command relationships, and distance to Baghdad, Col McFarland recognized this campaign warranted a different layout.

"93 Jump Package"

MWCS-38 established a critical retransmission site just north of the Kuwait-Iraq border that provided secure connectivity at Ali al-Salem Air Base and Camp Commando in Kuwait with extended tactical communications to forward operating sites. By forming a mobile team that deployed in Iraq, MWCS-38 established secure voice and data communications at multiple FOBs during offensive operations. Known as the "93 Jump Package," it comprised an AN/TSC-93C satellite package for Colonel McFarland's forward command element as it leapfrogged side-by-side with MWSG-37 to support 1st MarDiv.

Major Jeffrey P. Davis spearheaded most of the 93 Jump Package planning effort from a procedural air control perspective, coordinating with air bosses at the forward sites. Comprising a team of about 25 Marines with vehicle-mounted communications equipment, they staged initial operations at an-Numaniyah, a FOB, and eventually moved to ad-Diwaniyah, a FARP, until "nodal communications" were established south of Baghdad.[5]

which resided with 1st MarDiv at Camp Coyote, Kuwait. Three separate air support elements mirrored direct air support center functionality and were attached to General Mattis's forward division command post, 2d MEB's Task Force Tarawa, and United Kingdom forces (1st [UK] Armoured Division). Although not officially called direct air support centers to avoid confusion, each operated as such and was fully equipped, staffed, and capable of performing direct air support center functions. More specifically, controllers conducted procedural control of aircraft with pilot check-in and checkout protocols, thus providing route headings quickly. Additionally, McFarland stood-up a direct air support center (airborne) configured in an AN/UYQ-3A communications van and housed inside a

Col Ronnell R. McFarland, MACG-38's commanding officer, receives a field briefing from Maj Brent E. Sanders and the direct air support center crew while deploying forward from Kuwait.

Photo courtesy of Maj Jeffrey P. Davis

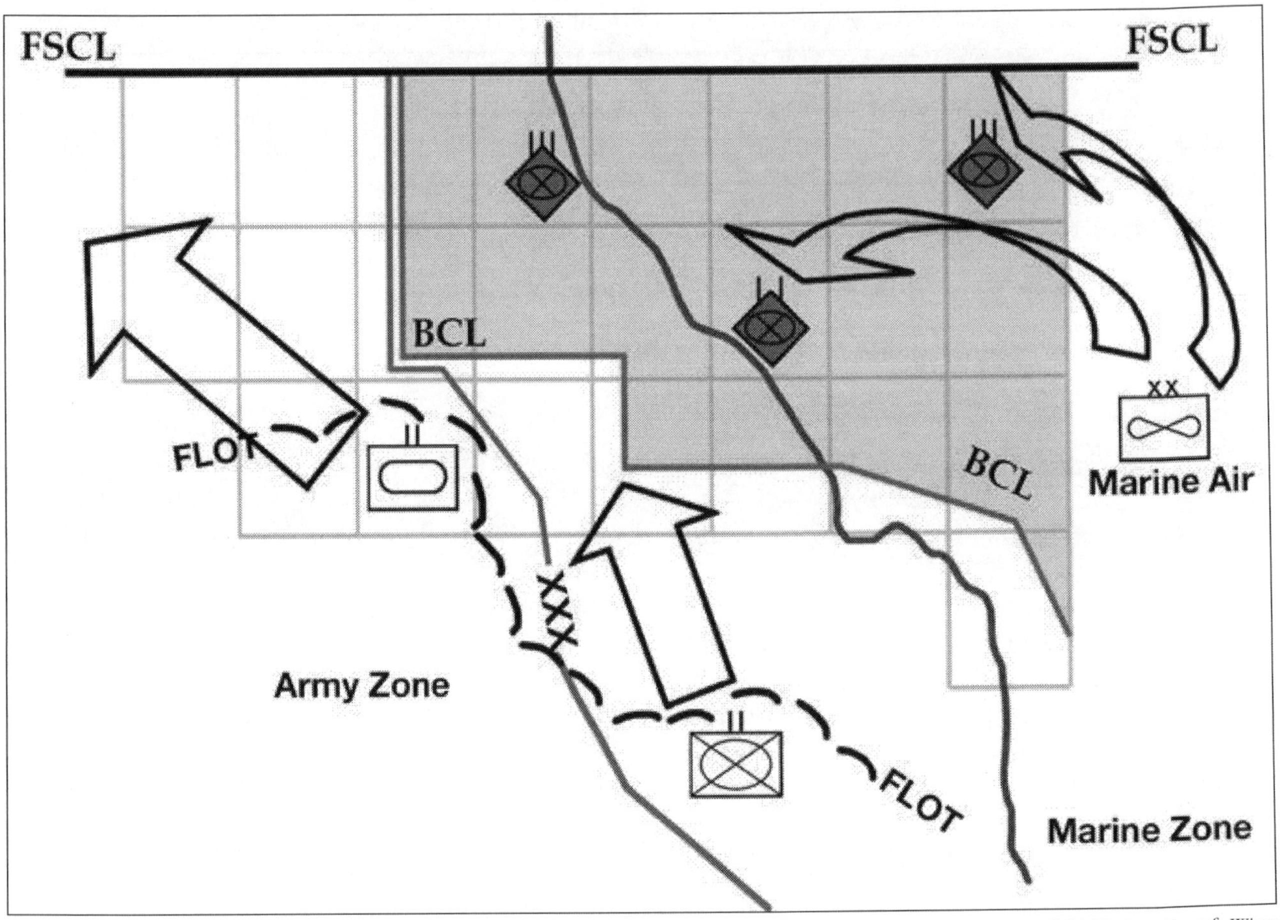

Map courtesy of 3d Marine Aircraft Wing

Col Ronnell R. McFarland used the battlefield coordination line (BCL) and added an extra layer of coordination for close air support between the fire support coordination line (FSCL) and the forward line of troops (FLOT), as shown.

specially equipped KC-130 Hercules. The airborne support center proved instrumental in maintaining contact with low-flying helicopters that often lost line-of-sight communications with the main direct air support center. Whether assigned to the main 1st MarDiv direct air support center in Kuwait, the airborne platform, or one of the mobile air support elements, tactical air controllers vectored aircraft and processed three types of air support requests—joint tactical air, air support, and medical and casualty evacuations.[6]

Colonel McFarland further decentralized air control capabilities to its lowest level to support the ground scheme of maneuver with the creation of four air support liaison teams to balance an anticipated heavy workload. Mounted in tactical vehicles, the teams moved independently with each of the three regimental combat teams—1st Marines, 5th Marines, and 7th Marines—with the 11th Marines artillery regiment. These support liaison teams did not have control capabilities, but forwarded air-to-ground information from the regimental level up to division.[7]

Second, clearly understanding that Marine Air Control Squadrons 1 and 2 would be significantly underutilized in this campaign—because the Iraqi antiair threat was already relatively low and air superiority would be quickly gained soon after the air offensive began—the colonel added a new role for the TACC. Air controllers assisted in strike coordination and reconnaissance missions, which were vital in searching for enemy activity and bombing enemy artillery targets in advance of the ground troops, particularly on the Marines' eastern flank. Both squadrons consolidated efforts to establish 24-hour operations that relied on the organic AN/TPS-59 and AN/TPS-63 early warning radar system located at Tactical Assembly Area Coyote in Kuwait. Operating beyond their typical antiair warfare function, air controllers worked directly with fighter and attack pilots, assisting in controlling "kill box" interdiction and close air support missions.[8] Through this combined effort, 3d MAW air controllers employed F/A-18D Hornets to search for targets of opportunity in advance of the ground combat element.

Kill boxes were distinct areas that identified grid coordinates to monitor combined arms fires in the close and deep fights, further distinguishing boundaries to minimize the potential for friendly fire incidents. The TACC typically managed only the air defense function, but operating under their new role, pilots had to check in with controllers to receive authorization to release bombs inside a kill box. Although kill boxes were not a new control measure—they had been used extensively during Desert Storm—McFarland enhanced their effectiveness by employing the TACC as the dedicated controlling agency for the kill box airspace.

Defining Airspace Coordination Measures

When MACG-38 received authorization to control its airspace, controllers used several measures to decentralize control of aircraft, including dictating altitude separation between fixed- and rotary-wing aircraft, using kill boxes to coordinate bombing runs, and publishing clear and concise pilot-controller protocols. These measures were designed to protect aircrews and ground forces from potential friendly fire incidents and to facilitate responsive weapons delivery on enemy targets. An otherwise oversaturated airspace could become dangerously unsafe and ripe for mishaps. In response, Colonel McFarland introduced and established another intermediary control point between the forward line of troops (FLOT) and the fire support coordination line (FSCL)—the battlefield coordination line (BCL). The BCL delineated airspace where Marine air controllers could manage on-call air support requests and targets of opportunity, but only by following strict protocols. The additional coordination line forced an extra layer of cautionary planning and demanded sharp focus to minimize the potential for friendly fire incidents.[9] The BCL, wedged between the FSCL and the FLOT, minimized breakdowns in communication and threats to friendly forces. The Marine Corps considered a kill box open, even if it was short of the FSCL. A controlling agency, however, such as a direct air support center or a forward air controller, had to confirm there were no friendly troops inside the kill box perimeter.[10]

The airspace between the FLOT and FSCL was off limits for close air support unless three criteria could be met: firm control of the airspace through an air controller; 100 percent positive identification of the target; and authorization to drop ordnance by the direct air support center.* By establishing the supplementary BCL, Marines retained an airspace that attracted Coalition aircraft in search of targets of opportunity in a controlled area where kill boxes could be opened more easily.[11]

As air operations intensified, the Marines earned a positive reputation for effectively directing aircraft to intended targets. One of the biggest compliments MACG-38 controllers received was neither spoken nor written, but rather repeatedly demonstrated when Air Force and Navy aircraft checked in with their respective service controllers, then requested to go to the Marine zone because the Marines were dropping bombs. As General Amos confirmed, "On any given day, around the clock, 24-hours a day, we were dropping ordnance. . . . The Marines were plugged into the Air Force and we plugged into them. General Moseley never asked for anything that I could not give him and I never asked him for anything he could not give the Marines."[12]

Leapfrogging with Marine Wing Support Group 37

Colonel Michael C. Anderson, MWSG-37's commanding officer, developed an aggressive approach for stretching air operations to accommodate the rapid advance to Baghdad. His group managed five Marine wing support squadrons (MWSSs), each assigned with a full range of aviation support functions, including airfield operational services, ordnance disposal, rescue, safety, refueling, tactical communications, weather forecasting, ground transportation, engineering, billeting, messing, medical services, and security.

Operating jointly with MACG-38 Marines,

---

*Detailed airspace management protocols allowed Marine pilots to release ordnance for close air support missions, whereas Air Force pilots typically conducted more strategic missions and were not as likely to fly under such conditions. A typical BCL extended 18.6 miles out from the FLOT—roughly the range of 105mm artillery. Air strikes short of this line were typically type 1, 2, or 3 close air support, calling for varying degrees of control.

Photo by LCpl Eric A. Archer. Defense Imagery 030422-M-DB300-002

Cpl Jeremy R. Jones (right) and LCpl John L. Dindlebeck, both assigned to the expeditionary airfield division of MWSS-272, disassemble aircraft runway matting in Kuwait that was no longer needed.

Photo by Sgt Giles M. Isham. Defense Imagery 030325-M-3368I-007

During Operation Iraqi Freedom, several Marine AH-1W Super Cobras land on a road near the Jalibah airstrip for fuel and munitions. The supplies were provided from a forward arming and refueling point operated by MWSS-373.

U.S. Marine Corps Air Bases, Forward Operating Bases, and Forward Arming and Refueling Points

Code Name	Location	Type
Ali al-Salem	Kuwait	Air base
al-Jaber	Kuwait	Air base
Sheikh Isa	Bahrain	Air base
MCAS Joe Foss	Kuwait	FOB/EAF
Turner Field	Kuwait	FARP
Astrodome	Kuwait	FARP
Busch Stadium	Safwan	FARP
Arlington	ar-Rumaylah	FARP
Riverfront	Jalibah	FOB/TLZ
Camden Yards	an-Nasiriyah	FARP
Fenway	Qalat Sikar	FARP/TLZ
PacBell	Highway 1	FARP
Wrigley	Highway 1	FARP/TLZ
QualCom	Shaykh Hantush	FARP/TLZ
Three Rivers	an-Numaniyah	FOB/TLZ
Ebbets Field	Sarabadi	FARP
Yankee	Salaman Pak East	FOB/TLZ
T. A. Wolf	North of Baghdad	FARP
	Highway 1	FARP
	Samarah	FARP
	Tikrit South	FARP/TLZ
	al-Kut	FOB/TLZ
	ad-Diwaniyah	FARP
	Tallil	FOB
	al-Hillah	FARP

MCAS = Marine Corps air station EAF = expeditionary airfield TLZ = tactical landing zone

MWSG-37 provided mobile aviation ground support and transported high-priority equipment, supplies, and personnel to various sites throughout Iraq.[13] As a critical component of the air strategy, General Amos depended on the support group to establish an aviation ground support plan to extend supply lines from stock points to remote air bases to provide adequate replenishment of fuel, ammunition, and repair parts for minor aircraft maintenance. Perhaps simple on paper, plan implementation required the intelligence and hard work of MWSG-37 Marines to establish the network of FOBs and FARPs, while constantly on the move supporting 3d MAW's mobility as a maneuver element.

While FOBs provided a more robust infrastructure for relatively long-term sustained air operations, FARPs were smaller and served as mobile air bases that could be relocated quickly based on the pace of the ground combat element. From the mobile bases close to the battlefield, aircraft could quickly land, refuel, rearm, and take off to reengage the enemy, thus maximizing their precious time-on-station, especially for the short range Cobras and Harriers. The mobile sites eliminated the need for much longer flights to and from bases in Kuwait and Saudi Arabia or aircraft carriers in the Gulf. Akin to a desert oasis, they offered temporary physical and psychological comfort. Aircrews could eat, rest, and gain situational awareness while their aircraft were refueled and rearmed.[14]

Named after baseball stadiums, the network of FOBs and FARPs from Kuwait to Tikrit was built by MWSG-37. Establishing FARPs and FOBs was not a new concept for the Marine Corps, but the number of sites and the magnitude of their employment during OIF greatly enhanced 3d MAW mobility. Additionally, the sites mitigated numerous aviation logistics issues associated with supporting aircraft in the battlefield far from supply and maintenance depots.

Without the mobility of MACG-38 and MWSG-37, 3d MAW aircraft could not have supported the ground scheme of maneuver or searched so extensively or effectively for enemy activity ahead of the ground forces. The combined efforts of these two groups enabled the air wing to operate as an independent maneuver element.

Chapter 5

Deploying Forward: Moving People, Planes, and Parts

The Logistician's Battle

A few days after the New Year 2003, General Amos received Deployment Order 164 directing 3d MAW's participation in OIF. Although the official deployment order was not received until 4 January, the planning effort had begun in earnest months earlier, especially for the logisticians who wrestled with a myriad of embarkation, transportation, and offload issues. Within a matter of weeks, a staggered arrival of more than 400 aircraft along with 15,000 air wing personnel converged at staging areas in Kuwait, totaling more than 80,000 I MEF Marines, sailors, and UK forces.

Colonel Juan G. Ayala, the wing's logistics director, described his thought process as he crafted an integrated logistical plan:

> Months before we arrived in Kuwait, we reviewed several spreadsheets and conducted a series of very detailed analyses. From the logistics perspective, our team repeatedly asked two fundamental questions. First, what would be the breakdown of aircraft, equipment, parts inventory, and personnel requiring transportation on U.S. Navy ships or by military or commercial air to the Kuwaiti naval base? Second, what would be the aircraft disposition once in theater?[1]

Col Juan G. Ayala served as the 3d MAW assistant chief of staff of logistics and oversaw one of the largest deployments during Operation Iraqi Freedom in terms of the movement of equipment, inventories, aircraft, and personnel.

Photo courtesy of Col Charles J. Quilter II

The result of these questions was several load and movement plans that relied on three modes of transportation—U.S. Navy amphibious ships, commercial and military aircraft, and individual squadron transatlantic ferry flights. Therefore, continuous liaison between I MEF and 3d MAW air-ground team planners for requesting Navy amphibious shipping and intercontinental Air Force and civilian flights was critical throughout the entire planning cycle.

Transportation by Sea

Ship movements from the East and West Coasts began in November 2002. Referred to as Amphibious Task Force–East (ATF-East) and Amphibious Task Force–West (ATF-West), embarkation was conducted at a number of military bases and ports. ATF-West departed Naval Base San Diego and Port Hueneme in California, while ATF-East deployed from Naval Station Norfolk and Naval Amphibious Base Little Creek in Virginia and from North Carolina's Port of Morehead City.

Many Marines assumed additional logistical roles that exceeded normal billet responsibilities to facilitate this massive deployment. For example, MAG-16's commanding officer, Colonel Stuart L. Knoll, filled a secondary billet as commander of ATF-West and oversaw the movement of multiple squadrons and units. His responsibilities included two MAG-13 Harrier squadrons—Marine Attack Squadrons 211 (VMA-211) and 311 (VMA-311) from Yuma, Arizona; two MAG-16 rotary-wing squadrons—Marine Heavy Helicopter Squadrons

165 (HMH-165) and 465 (HMH-465) from Miramar, California; and Combat Service Support Company 111 and Regimental Combat Team 1 (RCT-1) from Camp Pendleton and Twentynine Palms, California.[2]

Colonel Knoll and his small planning staff began the arduous loading process in late December 2002, finishing when MAG-16 departed San Diego on 17 January for a 44-day voyage to the northern Persian Gulf.[3] ATF-West included two amphibious assault ships (multipurpose)—the USS *Boxer* (LHD 4) and USS *Bonhomme Richard* (LHD 6)—and two amphibious transport docks—the USS *Cleveland* (LPD 7) and USS *Dubuque* (LPD 8). These four ships transported 3d MAW aircraft, including 24 AV-8Bs, 16 CH-53Es, 18 AH-1Ws, 9 UH-1Ns, and 12 CH-46s. The task force also included three dock landing ships—the USS *Anchorage* (LSD 36), USS *Comstock* (LSD 45), and USS *Pearl Harbor* (LSD 52).[4]

The AV-8B fleet demanded an additional planning effort because availability of naval amphibious shipping was at a premium. Flight deck constraints did not allow consolidation of the Harrier community on one amphibious ship, so the aircraft were split between ATF–West and ATF–East ships, with placement on the *Bonhomme Richard*, USS *Bataan* (LHD 5), USS *Tarawa* (LHA 1), and USS *Nassau* (LHA 4). The separation meant that most of the MAG-14 Harriers from Cherry Point, North Carolina—VMA-223 and VMA-542—departed from Morehead City on the *Bataan*, while one squadron—VMA-214—self-deployed on a transatlantic flight and flew directly to Kuwait. A remaining Harrier

ATF-West sailed from San Diego, California. Despite weather-related delays and a mechanical issue on board the USS Dubuque *(LPD 8), ATF-West arrived at Kuwait City on 23 February 2003.*
Illustration courtesy of 3d Marine Aircraft Wing

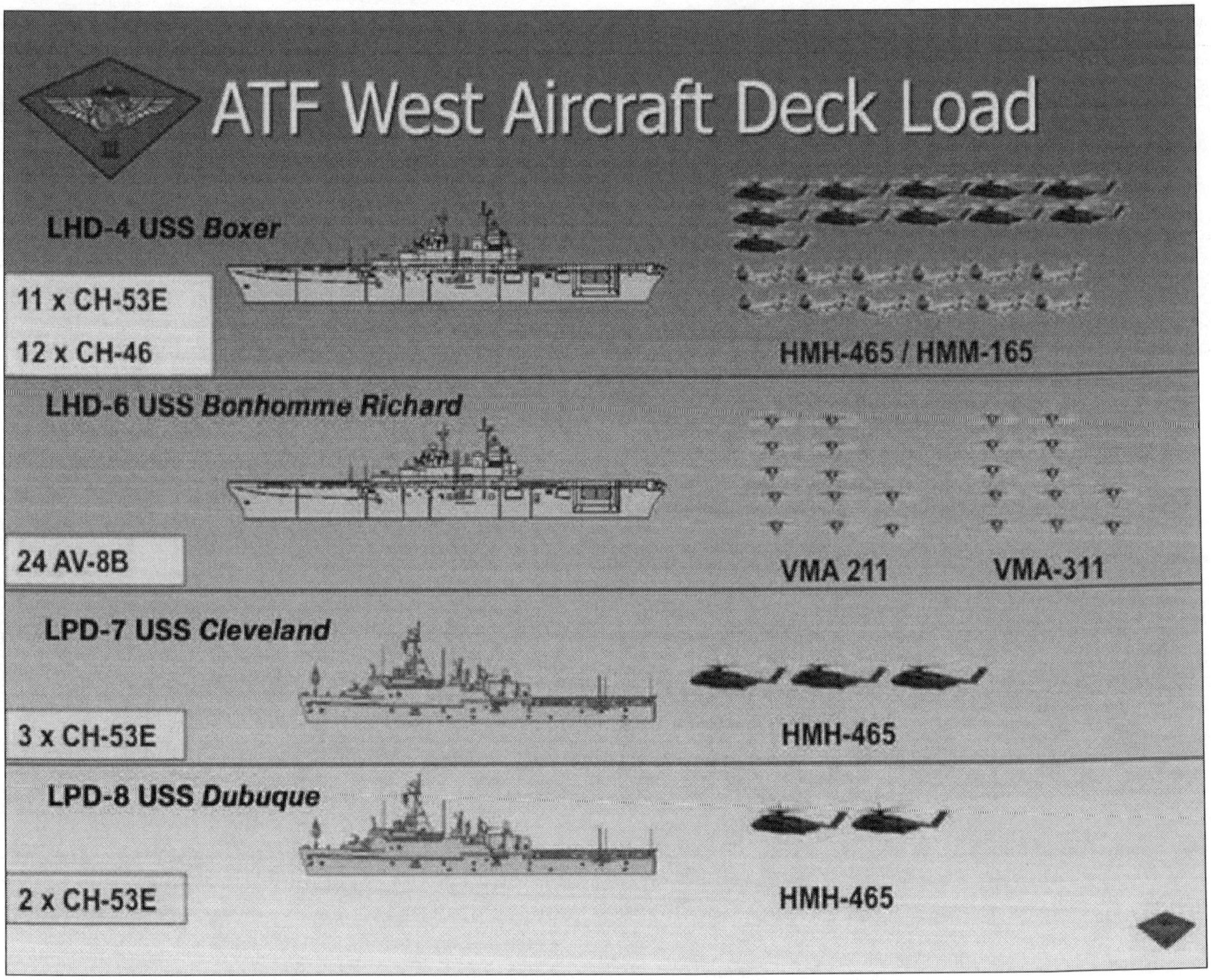

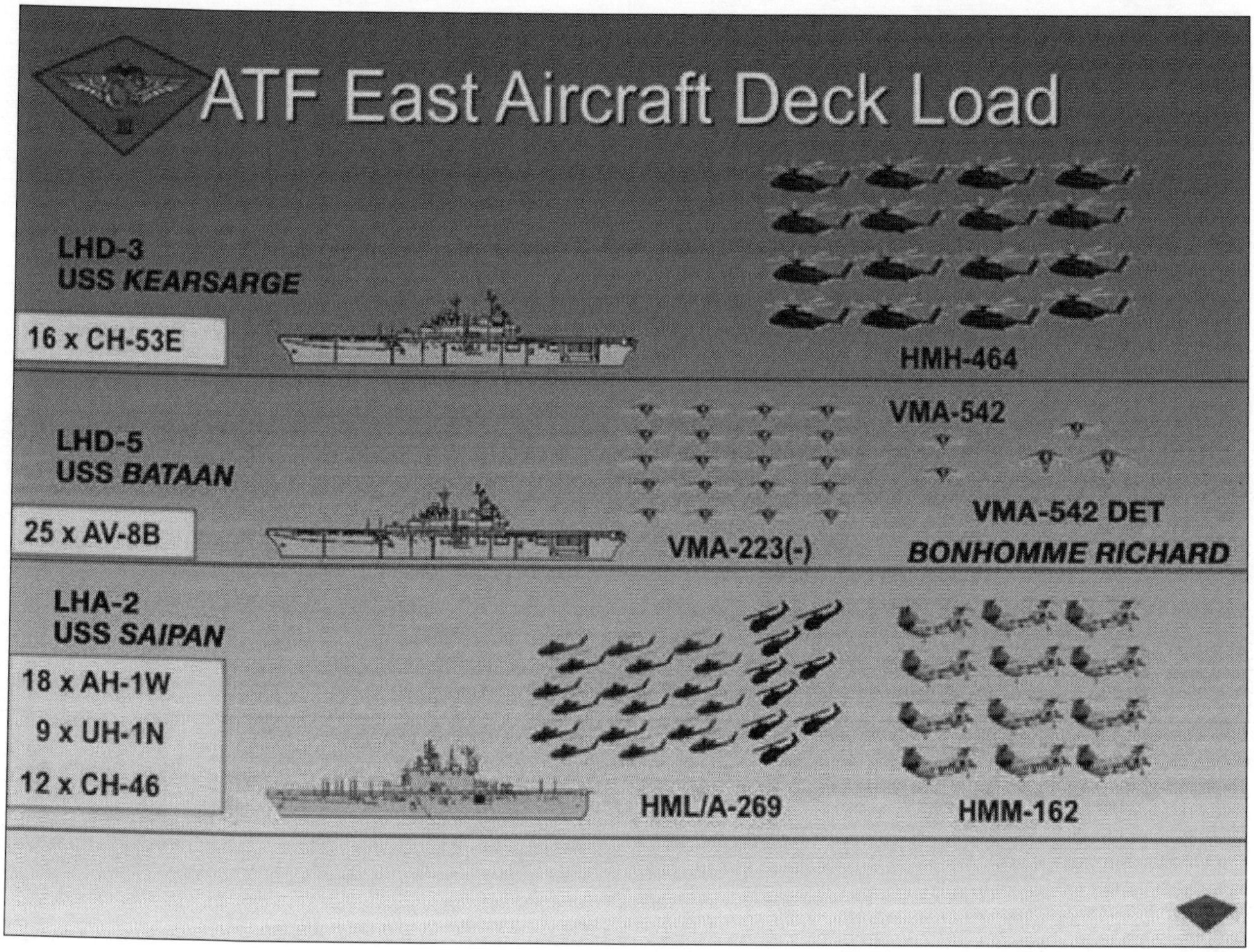

Illustration courtesy of 3d Marine Aircraft Wing

ATF-East departed Morehead City, North Carolina, transporting primarily 2d MEB aircraft, equipment, and personnel as an integral part of 3d MAW.

squadron already forward deployed—VMA-513—continued Operation Enduring Freedom missions in Afghanistan.[5]

Meanwhile, ATF-East transported 8,000 Marines and sailors from 2d MEB based at Camp Lejeune, North Carolina, with AV-8B Harriers and helicopters from MAG-29.[6] The task force, which departed by 17 January, consisted of two amphibious assault ships (multipurpose), the USS *Kearsarge* (LHD 3) and *Bataan*; an amphibious assault ship (general purpose), the USS *Saipan* (LHA 2); an amphibious transport dock, the USS *Ponce* (LPD 15); and three dock landing ships, the USS *Portland* (LSD 37), USS *Gunston Hall* (LSD 44), and USS *Ashland* (LSD 48).[7]

The 15th MEU entered the Persian Gulf on board the *Tarawa* with Marine Medium Helicopter Squadron 161 (HMM-161). A sister squadron, HMM-263, sailed with the 24th MEU (SOC) on the *Nassau*.

Transportation by Military and Commercial Air

Personnel, aircraft, equipment, and cargo not moved by the Navy arrived in Kuwait on board commercial and military flights. The National Command Authority* activated the Civil Reserve Air Fleet to speed the flow of forces to the region. The Reserve Air Fleet consists of aircraft contractually obligated by their airlines to provide transport services on an as-needed basis. Once called up, the airlines usually had between 24 and 48 hours to provide the aircraft, which they operated and maintained; however, missions were controlled

*Directions for military operations come from the National Command Authority, a term used to collectively describe the president and the secretary of defense. The president, as commander in chief of the Armed Forces, is the ultimate authority. The Office of the Secretary of Defense carries out the secretary's policies by tasking the military departments, the chairman of the Joint Chiefs of Staff, and the unified commands.

Photo courtesy of Col Robert E. Milstead Jr.

Referred to as the "Magnificent Seven," ATF-East sailed from the port of Morehead City, North Carolina, and passed through the Suez Canal, transporting helicopters and AV-8B Harriers from the 2d MEB.

Origin of the Maritime Prepositioning Force

After the Vietnam War, economic concerns over high defense costs and budgets threatened the Marine Corps' ability to deploy combat-ready forces where needed, when needed, sometimes on short notice. Ongoing strategic and operational commitments, along with the Corps' resolve to carry out its mandated expeditionary mission, resulted in analyses of various basing and deployment patterns that ultimately resulted in the Maritime Prepositioning Ship program and its integral Maritime Prepositioning Force (MPF).

Introduced in late 1979, MPF provided forward-deployed resources for three Marine amphibious brigades on 15 preloaded ships organized into three squadrons. The ships were located at ports around the world and configured to support contingency plans for the rapid deployment and movement of Marines during the initial stages of an operation. Each squadron contained enough supplies and equipment to support a 15,000-person MAGTF for 30 days.[8] The concept gave the U.S. military and the Marine Corps a significant new dimension in mobility, sustainability, and global response capabilities. The program also facilitated brigade-size airlifts to a potential crisis area. At predetermined locations, units would be reunited with their ships and essential equipment and supplies.

In OIF, the Marine Corps relied heavily on the Navy's amphibious assets and capabilities to coordinate the movement of aircraft, equipment, supplies, and a 30-million-pound ordnance cache. The 3d MAW planners used 11 MPF ships, transporting 98 percent of its assigned equipment and 100 percent of its aviation ground support equipment in standard international ocean containers.[9] Additionally, the Marine Corps contracted three ammunition ships and four Fast Sealift Ships to transport helicopters from several squadrons.[10]

Photo by LCpl Ronoldson G. Slim.
Defense Imagery 030211-M-SG577-061

Marines from MWSS-371 board a United Airlines Boeing 747 commercial airliner at Marine Corps Air Station Yuma, Arizona, for deployment in support of Operation Iraqi Freedom.

by the Air Force's Air Mobility Command. Marines boarded commercial jets and ultimately arrived in theater at staging areas for final assignment, while remaining squadron support personnel flew on military transport aircraft or on board regular commercial airliners.

Offload at Base Camps

Colonel Ayala and his staff oversaw the reception, staging, onward movement, and integration phases, which constituted the initial part of the deployment. The advance party arrived in Kuwait in November 2002 to prepare for the massive influx of the air wing Marines, tactical aircraft, equipment, and parts. The 135-member team initiated several aggressive reconstruction projects that included improving existing facilities, repairing runways, and expanding parking aprons at al-Jaber and Ali al-Salem airfields in Kuwait. The equipment offload effort proved to be just as challenging as the embarkation exercise, especially when naval ships and military and commercial aircraft converged in Kuwait, causing great demand for ground transportation and heavy equipment assets.

Marine Capt Kevin C. Rosen (right) provides 3d MAW commanders MajGen James F. Amos (second from left) and BGen Terry G. Robling with a tour of the staging area for the maritime prepositioning ship offload site near Shuaiba Port, Kuwait.

Photo by LCpl Christopher H. Fitzgerald. Defense Imagery 030128-M-YA501-004

Embarkation Metrics

Just as the Marine Corps relied on the Navy as its amphibious partner, the Air Force also provided tremendous airlift capacity for 3d MAW equipment, cargo, and personnel. During the movement to Kuwait, the Air Mobility Command moved more than 1,979 tons of cargo and transported 3,103 military personnel to the CentCom area of responsibility. Cargo and passengers were flown on Boeing C-17 Globemaster III, Lockheed C-5 Galaxy, Lockheed C-141 Starlifter, and C-130 Hercules transports. The C-5 was capable of carrying 36 standard pallets and up to 81 troops simultaneously.

The 3d MAW offloaded more than 7,216 tons of cargo from the MPF and Fast Sealift Ships, which equated to approximately 87,700 inventory line items and 1,409 containers.[11] Additionally, the wing offloaded three ammunition ships that carried 29.1 million pounds of aviation ordnance, which was one of the largest stockpiles ever recorded.[12]

After naval ships anchored in the port of Kuwait, 3d MAW offloaded two MPF squadrons—11 transports and 4 Fast Sealift Ships—that ferried aircraft from several helicopter squadrons. The offload effort demanded a continuous 24-hour shift for 45 consecutive days without an injury to any Marine or sailor.[13]

Photo courtesy of Maj Jeffrey P. Davis

This view of the tent city at Ahmed al-Jaber Air Base provides a glimpse of the living conditions for 3d MAW Marines in Kuwait.

Colonel Ayala exercised a management-by-walking-around approach, supervising the logistical plan as it unfolded and making key decisions seemingly on an hourly basis. Most of the 3d MAW personnel lived at the two Kuwaiti air bases—Ahmed al-Jaber and Ali al-Salem. Each base contained more than 400 tents and offered chow halls, restrooms, and shower facilities. At one point during the construction of the tent cities, Ayala expressed his frustration when he was unable to get the equipment, supplies, and personnel to complete the mission: "We needed guys to drive our vehicles. Instead, I was getting a lot of planners, and liaison officers, and colonels, and one particular day, I was very upset; I called back to my chief of staff and relayed a message. I pled, 'Send me cots, because my people are sleeping on the deck. Send me corporals, but don't send any more d——n colonels!'"[14]

On 3 January 2003, 3d MAW assistant wing

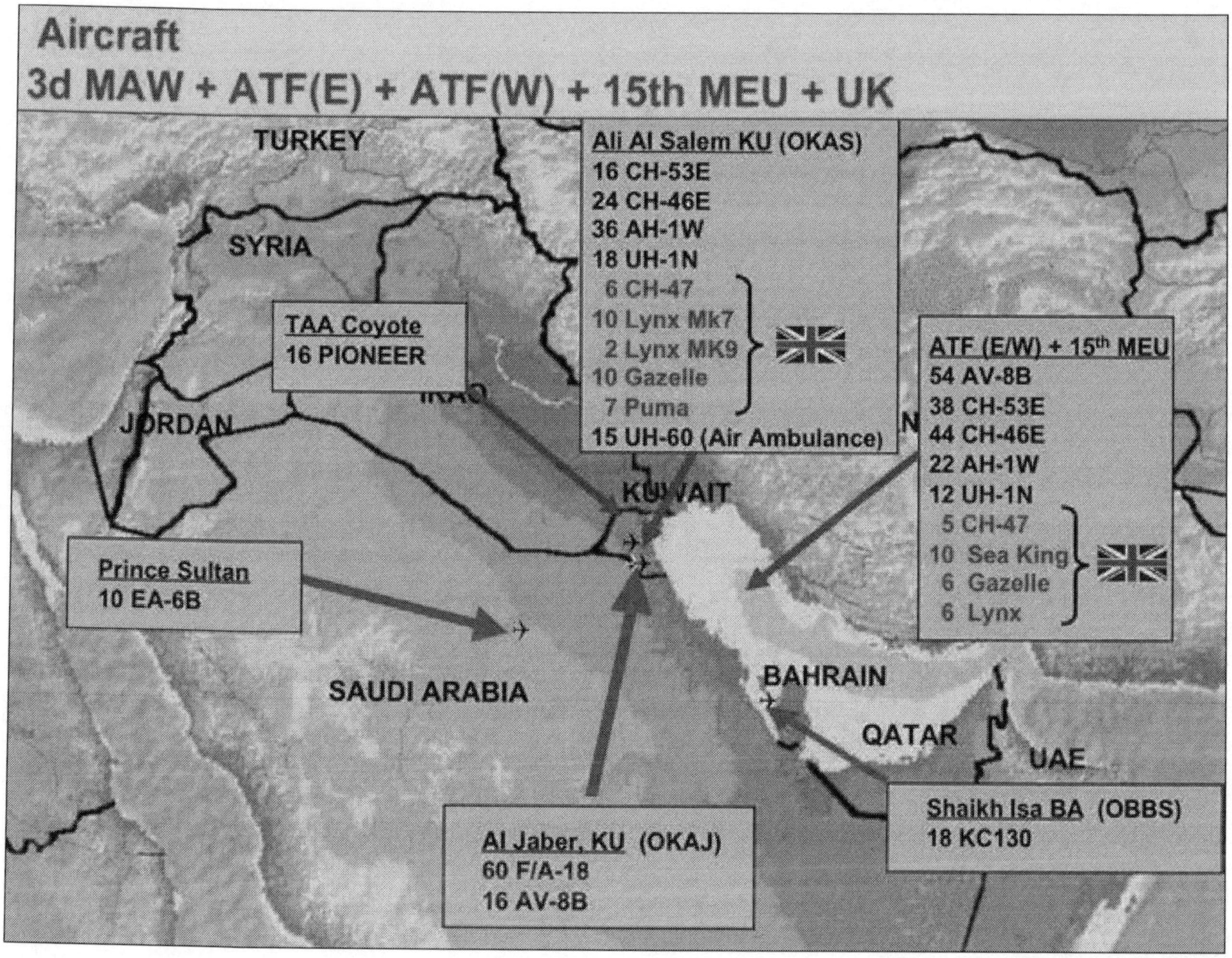

Map courtesy of 3d Marine Aircraft Wing

Aircraft disposition when Operation Iraqi Freedom began and before the 3d MAW established forward operating bases deep inside Iraq.

commander, General Robling, joined the advance team and monitored the upcoming staggered touchdown of more than 400 Marine aircraft and the offload of the largest debarkation effort, in terms of aircraft and personnel, since Operation Desert Shield.[15] After the offload was completed in mid-February, Robling recalled, "The planners allotted a 60- to 90-day flow for the time-phased force movement of cargo and personnel. However, 3d MAW logisticians exceeded our expectations. They received and unloaded our ships in about 45 days. It was unprecedented."[16]

Fighter/Attack and Transport Aircraft

When MAG-11's commanding officer, Colonel Randolph D. Alles, arrived at Ahmed al-Jaber Air Base on 21 January, his advance party had only a few days' wait until the main body arrived, an assortment of aircraft that collectively could fulfill five of six Marine aviation functions—offensive air, antiair warfare, assault support, aerial reconnaissance, and electronic warfare. His group ultimately totaled 60 F/A-18s from five Marine fighter/attack squadrons and 24 KC-130s from Marine Aerial Refueler Transport Squadron 352 (VMGR-352) with augmentation from two reserve detachments, VMGR-234 and VMGR-452. Forming a Hercules air group, the multimission KC-130 conducted a variety of missions, including tactical aerial refueling, assault air transport, and a rotation to support a direct air support center (airborne).[17] Additionally, two squadrons of EA-6Bs were based at Prince Sultan Air Base in Saudi Arabia, where, as a national asset, they received joint antiair warfare tasking. Marine Aviation Logistics Squadron 11 (MALS-11), commanded by Lieutenant Colonel Shaugnessy A. Reynolds, along with detachments from 2d MAW—MALS-14 and MALS-

Revisiting the "Harrier Carrier" Concept

Twelve years after Operation Desert Storm validated the "Harrier carrier" concept from the USS *Nassau* in the Persian Gulf, 3d MAW reintroduced and refined it for OIF. Although the initial idea of AV-8B Harrier air operations and maintenance activities on board Navy amphibious ships stirred debate among Navy and Marine Corps commands, its execution during Desert Storm proved successful, with the *Nassau* averaging more than 60 Harrier sorties per day.

Regardless of this success, lingering issues from the 1991 operation required planners to resolve problems and overcome challenges regarding airspace constraints, increased physical separation between amphibious ships, aircraft consolidation to a single carrier platform, and ordnance resupply accessibility. As part of the enhancement initiative, the Navy's joint intelligence center expanded its focus to include expeditionary strike operations as part of its normal collection effort and offered more detailed weather forecasting reports that could affect launch and recovery decisions.

Improved weather briefings were invaluable during combat operations. As demonstrated over five days, the Marines cancelled 103 sorties because of adverse real-time data provided by forecasters. This information was critical data that contributed to zero weather-related aircraft mishaps or diversions during the deployment.[18]

When the USS *Bataan* launched its first series of AV-8Bs to support OIF on 20 March,

Photo courtesy of Col Charles J Quilter II

Capt Jennifer M. Dolan flew one of the first AV-8B Harrier missions in support of Operation Iraqi Freedom from the USS Bataan *(LHD 5).*

Captain Jennifer M. Dolan was a wingman during one of those early flights. She recalled the conditions that evening: "I was flying with the operations officer and the weather was really crappy. In fact, we sat on the deck ready to go for about an hour and the ship almost did not let us launch. We needed a mile-and-a-quarter to a mile-and-a-half [of visibility] to be legal to takeoff from the boat, which is still pretty poor considering that you would return and land on it. We eventually launched with two miles visibility."[19] One day later, a MAG-13 aircraft delivered the group's first weapon on target.[20]

Photo courtesy of Col Charles J Quilter II

Maintenance and flight line crews oversee Marine AV-8B Harriers on board the USS Bonhomme Richard *(LHD 6).*

31—joined the group roster, providing essential logistical and maintenance support activities for all 3d MAW fixed-wing aircraft. Colonel Alles's composite fleet was spread out across three bases in Kuwait, Saudi Arabia, and Bahrain with aircraft immediately flying round-the-clock missions to support Operation Southern Watch. This was only the second time since the Vietnam War that MAG-11 deployed forward as a full group.[21]

Sea-Based Operations

The MAG-13 commanding officer, Colonel Mark R. Savarese, faced a shortage of available land-based air facilities in Kuwait when planning to deploy 64 AV-8B Harriers, so 3d MAW partnered with the Navy and revitalized the "Harrier carrier" concept from the Gulf War. Colonel Savarese merged three West Coast Harrier squadrons based at Yuma—VMA-211, VMA-214, and VMA-311—with two East Coast squadrons—VMA-223 and VMA-542—from 2d MAW. When all squadrons arrived in theater, the air group totaled 64 Harriers with MALS-13 providing logistical and maintenance readiness.[22] Just like MAG-11, MAG-13 aircrews immediately began flight operations supporting Operation Southern Watch, which allowed aircrews time to familiarize themselves with the new area of responsibility while gaining expertise using new targeting and navigational pods.

Helicopters

To better support the ground scheme of maneuver, 3d MAW staged most of its utility, transport, and attack helicopters at FOBs near division operations. By moving with RCT-1, RCT-5, and RCT-7, helicopter aircrews frequently met with ground units, conducting face-to-face liaisons before executing missions during the march to Baghdad. Not surprisingly, the arrangement improved mission responsiveness support and situational awareness, as well as air-ground team camaraderie—further reinforcing General Amos's value on building relationships.

MAG-16's commanding officer, Colonel Knoll, initially launched CH-46s and CH-53s from two locations: from al-Jaber Air Base in Kuwait and from the flight deck of the *Boxer*. When MWSG-37 and MACG-38 established an FOB in Jalibah, Iraq, MAG-16 consolidated most of its operations there yet retained one squadron at Ali al-Salem in Kuwait.[23]

MAG-29's commanding officer, Colonel Robert E. Milstead Jr., traveled on board ATF–East with the Corps' only composite helicopter group, including CH-46 Sea Knights, CH-53 Sea Stallions, AH-1W Cobras, and UH-1N Twin Hueys from New River, North Carolina, and one squadron on board a Fast Sealift Ship, the USNS *Regulus* (T-AKR-292). Originally assigned as the aviation combat element for the 2d MEB, 3d MAW assumed operational control of MAG-29 aviation assets during combat operations.[24] By the first week of April, MAG-29 moved most of its helicopters ashore at Jalibah.

The commanding officer of MAG-39, Colonel Richard W. Spencer, managed AH-1Ws, UH-1Ns, and CH-46s at Ali al-Salem Air Base. Although the group's focus was on supporting Marines on the ground, it also oversaw and transported embedded media representatives throughout the I MEF area of operations.[25]

Hercules Aircraft Group

By February 2003, General Amos realized 3d MAW needed to stage a detachment of KC-130 Hercules transport tankers closer to 1st MarDiv operations. Although Sheikh Isa Air Base in Bahrain was home for the fleet of 24 KC-130s that formed the Hercules Aircraft Group, the general directed the construction of an expeditionary airfield near the Iraqi border in northern Kuwait. It would provide improved on-call response time for assault support, cargo, or refueling missions by operating closer to the division's regimental combat teams.

The austere field on the northern edge of Kuwait, FOB Joe Foss, offered an ideal location for a KC-130 detachment and as an additional forward site for helicopters and Harriers. Built by Marine Corps combat engineers and MWSG-37, their joint efforts took about three weeks to complete. During the project, additional logistical requirements were identified for airfield AM-2 matting, lighting, and runway repair parts.

Rear Air Operations

Despite personal desires, not every Marine who wanted to deploy to the Persian Gulf region could do so because air operations at Miramar, Cherry Point, Yuma, New River, and Beaufort did not cease. Although activities slowed initially, those Marines who sustained air operations stateside soon discovered a sense of urgency in supporting forward operations. At Miramar, initial plans to

Honoring a Legend

Located at Camp Coyote in northern Kuwait, 3d MAW named its FOB in honor of World War II Marine ace, Major Joseph J. Foss, who had recently passed away. Born on 17 April 1915 near Sioux Falls, South Dakota, Joe Foss received his naval aviator wings and a Marine Corps officer commission on 31 March 1940. One day later, he was called to active duty and then went on to pilot training in Pensacola, Florida. During his combat tour in the Pacific, Foss shot down 26 Japanese airplanes and was a Medal of Honor recipient. He left the Corps as a major and was appointed a lieutenant colonel in the South Dakota Air National Guard, which he founded and led. He later advanced to the rank of brigadier general in the Air Guard and became a two-term South Dakota governor.

The idea to build the airfield originated in February when General Amos and General Usher, commanding general of 1st FSSG, toured an ammunition storage site located near newly constructed medical facilities a few miles north of Camp Commando in Kuwait. Usher noted its proximity to the medical facility and suggested that 3d MAW build a runway to accommodate ammunition logistical resupplies and medical evacuations, eliminating the 35-minute drive to Ali al-Salem Air Base. Amos agreed and asked the operations officer, Colonel Miclot, about the feasibility of this proposal. After a few phone calls and discussions about prevailing winds, runway length, and nearby power lines, Miclot replied, "Sir, how big you want it?"[26]

Marine Corps KC-130s flew many missions from the dual runways at FOB Joe Foss in Kuwait, transporting personnel, equipment, and cargo to support Operation Iraqi Freedom.

Photo by LCpl Jonathan T. Spencer. Defense Imagery 030401-M-2900S-008

Photo courtesy of Col Charles J. Quilter II

Although remote and at times swallowed by dust storms, FOB Joe Foss was also used heavily by helicopter aircrews, transporting ammunition, Marines, and medical evacuees to a nearby medical facility during the march to Baghdad.

Teaming with Colonel Anderson's MWSG-37, 3d MAW began making plans for a dirt runway 5,000 feet long and 200 feet wide with parking aprons. Miclot made a few more phone calls back to al-Jaber, which led to a discussion about a potentiometer—an instrument to measure ground density. Because one was not readily available, the team used Marine Corps ingenuity and asked General Amos to dig the heel of his boot into the sand. As reported, the indentation of his boot heel was less than an inch, which seemed to answer the density question. So the digging began.

MWSS-272 began building the expeditionary airfield on 9 February 2003, including a 320,000-gallon tactical airfield fuel dispensing system, a 300-man base camp with billeting and concrete force protection bunkers, heads, operations and maintenance tents to support the population of the camp, a FARP, three tactical helicopter landing pads using AM-2 matting, two 6,000-foot runways, and five ramps for KC-130s. The runways were constructed of dirt and a compaction material of water and lime called "getch."

A taxiway between the two runways answered parking space concerns, and the dual runways allowed one to be operational while the other was graded for use the next day. FOB Joe Foss—the largest expeditionary airfield constructed since World War II—opened quietly without fanfare on 6 March when the first KC-130 landed,[27] but, as with its namesake, the airfield was a major contributor to Marine Corps air operations.[28]

reduce air station operations changed to 24-hour support for two unplanned projects—one for a local hospital and the other for the North American Aerospace Defense Command.* Many reservists backfilled active duty billets and undertook important roles supporting OEF and OIF.

At stateside air bases, Marines sustained air operations by keeping runways open, providing base security, expediting requests from deployed units, and training aircrews. In Southern California, 3d MAW fixed- and rotary-wing training squadrons continued courses for pilots and aircrews at Miramar with Marine Fighter Attack Training Squadron 101, Camp Pendleton's Marine Medium Helicopter Training Squadron 164, and Marine Helicopter Training Squadron 303. At Naval Air Station Pensacola, Florida, flight instructors maintained training syllabuses and graduated naval aviators to extend the pilot pipeline for future unit rotations in the Middle East or wherever needed. Tactical weapons training at the Marine Aviation Weapons and Tactics Squadron 1 (MAWTS-1) in Yuma temporarily shut down its exercises because most instructors joined the 3d MAW staff, offering an exceptional skillset and wealth of knowledge on the battlefield in either planning and support roles or forward as liaison officers. Meanwhile, aviation logistics and maintenance specialists at Headquarters Marine Corps hounded procurement managers about the status of critical aircraft components and inventory shortages while monitoring Navy aviation supply channels and tracking shipments to Kuwait, Bahrain, and Saudi Arabia.

Colonel Robert W. Hillery, I MEF chief of staff (rear) at Camp Pendleton, focused on personnel and equipment deficiencies in preparation for the massive January deployment and throughout sustained combat operations. During the war, he oversaw approximately 10,000 I MEF Marines located at Camp Pendleton, Miramar, and Twentynine Palms and in Yuma. This included dealing with casualty notification, chaplain support, and a key volunteer network. As an F/A-18 Hornet pilot during the Gulf War, Colonel Hillery found it difficult to remain behind and watch fellow Marines; longtime aviator friends; and even family members, such as his son-in-law Major James B. Hanlon, a Huey pilot with the 15th MEU, prepare for war and deploy forward. The colonel also awaited his son's graduation from The Basic School officer training at Quantico, Virginia, fully realizing that if hostilities continued beyond Pentagon projections, many newly graduated lieutenants would join the deployment cycle.*

Photo courtesy of Col Robert W. Hillery

Maj James B. Hanlon and his young son, James Connor, stroll down the pier together just before the USS Tarawa *(LHA 1) departed for the Middle East. Marines from the 15th MEU bid farewell to their families in January 2003, unsure of when they might return home.*

In the final days before I MEF departed, Hillery recalled the Marines' tremendously high morale and positive sentiment:

> Quite honestly, it seemed a little undersized for what we were planning. Regardless, there was no doubt in my mind that anybody leaving the United States was going in[to] harm's way and I felt reassured knowing first-hand the level and intensity of pilot training and fully believing our Marines were well prepared. As a Marine and parent, I was emotional and, for the first time ever, I understood how difficult it was on families, given the uncertainty and waiting. I definitely would have preferred deploying forward. It was hard, really hard.[29]

*The North American Aerospace Defense Command (often referred to as NORAD) is a binational U.S. and Canadian organization charged with the missions of aerospace warning and control for North America.

*His son, 2dLt Robert J. Hillery, graduated from The Basic School in March 2004 and deployed to Iraq five months later with Truck Company Battalion, 1st MarDiv.

Part II

Readiness and Shaping the Battlespace

Chapter 6
Briefing the Air Campaign

Rehearsal Drills

In the months preceding hostilities with Saddam Hussein's regime, General Amos discussed the 3d MAW air campaign with the air-ground team, joint agencies, and Coalition partners during numerous command briefings. But one of the general's most effective means of communication was through his cadre of trusted liaison officers, who endorsed the commander's intent by their forward presence among joint staffs, planning cells, and regimental combat teams. These Marines reinforced the aviation strategy on a daily basis, influencing decision making and executing the air wing's campaign as Generals Amos and his deputy, General Robling, had intended.

As the wing began its phased departure from continental U.S. ports and airports to the Middle East, planning efforts shifted to conducting a series of well-orchestrated "rehearsal-of-concept" drills. On 6 January 2003, 3d MAW conducted the first such exercise at its operations center at Marine Corps Air Station Miramar with senior commanders from I MEF, 1st MarDiv, and 3d MAW in attendance. It coincided with the start of the wing's phased deployment to Kuwait by ship and

Gen Amos's 3d MAW commanders and liaison officers attend a rehearsal of-concept drill that integrated a visual display of units and simulated movements in preparation for pending hostilities in Iraq.

Photo courtesy of Col Michael D. Visconage

Photo courtesy of 1st Marine Division

The I MEF drill on 27 February visually displayed the aviation ground scheme of maneuver and the sequence of events for the initial days of the war, later referred to as the Opening Gambit.

air. Several of the commanders in attendance had recently returned from Kuwait, only to go back a few days later. The drill outlined the Marine aviation strategy, specifically describing how the aviation combat element would support the ground combat element and also maneuver in advance to attack deep targets ahead of ground forces.[1]

On 7 February, 1st MarDiv hosted a major rehearsal drill in the sand at its Camp Matilda, Kuwait, headquarters for the I MEF commander—General Conway—and his staff. The practice replicated the first 96 hours of OPLAN 1003V. General Mattis had directed the Marines to visually display key terrain features, participating units, and the scheme of maneuver. They produced a vast model of the obstacles in northern Kuwait, including its oil fields, rivers, ports, and roads on the Iraqi side of the border. Framed on one side by four seven-ton trucks forming a makeshift grandstand overlooking an area the size of a football field, the exercise lasted for three hours, with U.S. Marines, Royal Marines, and British soldiers, all in colored jerseys representing their units, literally walking through the moves that would occur during the first few days of the war.[2] The next day, I MEF participated in another round of a Coalition forces land component commander–sponsored exercise, Lucky Warrior 03-2, intended as the final test of command and control between the Army and Marine Corps. The five-day exercise was a dress rehearsal for war. It encompassed final OPLAN 1003V revisions that centered on a two-prong attack from Kuwait to Baghdad, rather than a single avenue of approach. Additional rehearsal drills were conducted throughout February and March.[3]

On 27 February, Lieutenant Colonel Michael S. Groen, a 1st MarDiv intelligence officer, spearheaded an event in which I MEF commanders held a second "jersey drill" to visually display a terrain model of the air-ground team's scheme of maneuver during the opening gambit. This rally was held east of Life Support Area Matilda with Generals Conway, Amos, and Mattis; Major General Keith J. Stalder; British Major General Robin V. Brims; Brigadier General Richard Natonski; Brigadier General Edward Usher; and Navy Rear Admiral Charles R. Kubic in attendance.[4] The 1st MarDiv engineers constructed a desert plain with bulldozers, including a multitiered amphitheater approximately 100-meters long with an angled surface for better viewing from the rear. Marines and soldiers dressed in colored sports jerseys physically walked through the actions that their units would take after crossing the line of departure.

After the rehearsal drills, General Amos made squadron visits to speak with Marines prior to hostilities. On 4 March, he visited the *Bataan* and emphasized that the war would soon start. In speaking, he reiterated several important themes: the need to watch out for one another, the just cause of the Coalition's impending actions, the need to limit collateral damage, and the belief that many Iraqi soldiers would rather surrender than fight.[5] The next day, he went on board the *Bonhomme Richard* and met with VMA-311 to brief pilots on the order of battle. Over two days, he flew to all six large-deck amphibious ships to ensure the commanders and their Marines fully understood the scheme of maneuver. A few days later, the 3d MAW judge advocate general followed-up with a formal slide show briefing and scenario training on the rules of engagement. Face-to-face sessions afforded aircrews a chance to get specific questions answered on target engagement and laser usage, among other points.[6]

Contributing to the precampaign briefing schedule, Colonel McFarland, the commanding officer of MACG-38, hosted a group commanders' conference at al-Jaber Air Base on 12 March, just eight days before combat operations began. All of the commanding officers, executive officers, operations officers, and sergeants major attended this final opportunity to review the concept of employment and to address unresolved aviation issues in the group forum. The conference featured opening comments by General Amos, who expressed his concerns and thoughts regarding potential friction within the Marine air command and control system. Feedback reported from squadron commanders indicated that the sessions were informative and thought-provoking, resulting in a positive prelaunch feeling of confidence about the upcoming operation.[7]

Incorporating Advanced Communications

There is little question that planners and commanders benefited from technological advancements in digital and satellite communications, yet enhanced capabilities generated installation and security issues for 3d MAW's assistant chief of staff for communications, Lieutenant Colonel David P. Olszowy. The colonel's staff erected transportable AN/TRC-170 microwave radio terminals that provided both secure and nonsecure connectivity between major nodes of the communications network. The limited bandwidth coupled with a high number of users requesting system access generated significant slowdowns on network response times. To improve this, wing communications installed a global broadcasting system that allowed staff to download sizeable, secure files by way of satellite, to communicate via Iridium satellite phones, and to trace ground and aviation forces through blue force tracker systems. Additionally, commands used video teleconferencing, a common civilian corporate communications tool. Although a rather sophisticated option in the battlefield, it introduced a new way of doing business for the Marine Corps, particularly for commanders and planning cells that were often several hundred miles apart.[8]

Digital and satellite communications offered a more expeditious means of exchanging information than in previous conflicts, but the advantages became a juggling act for those Marines who had to balance communications with system security risks. Colonel George J. Allen, I MEF communications director, addressed those concerns and instituted strict communication and security guidelines. By prioritizing and limiting unnecessary user access, key staff had full bandwidth capabilities to send and receive critical messages and data. A planning team oversaw and filtered all communications in the expeditionary force area of operations in what was referred to as the "River City Plan." It categorized users as Alpha, Bravo, or Charlie and facilitated information flows at al-Jaber Air Base and later at FOB Jalibah. By March, the effort shifted and the focus was to install, operate, and maintain a robust, reliable network inside the tactical air command center, where Generals Amos and Robling would supervise the air campaign when not deployed forward.

As hostilities loomed within 48 hours, General Amos directed the wing staff to secure nonessential communications.

> We stopped unclassified e-mail and conveyed [that] there would be no more e-mail or network access. I believe Brigadier General Robling, the chief of staff, the chaplain, and myself were the only ones who even had secure network capability. I spoke to my wife briefly and told her to brace herself and prepare the other spouses through the key volunteer network for a period of silence. We turned off communications capability because we knew when it was going to happen

Fielding New Communications Technologies

The I MEF staff and its major subordinate commands fielded several new communications technologies that significantly revamped the way the commanders exchanged information on the battlefield. One of the most lauded advancements was a piece of security equipment—the AN/TSC-154 Secure Mobile Antijam Reliable Tactical-Terminal mounted on High-Mobility Multipurpose Wheeled Vehicles (HMMWVs, usually referred to as "Humvees")—that allowed secure phone connectivity with the division commander within 10–15 minutes. Equally impressive was the AN/TSQ-222A tactical data network gateway and the AN/TSG-227 digital technical control facility with its inherent multiplexing capability, thus enabling I MEF to establish a complex and extensive communication architecture, unlike earlier Marine Corps combat deployments.

From the start of air combat operations on 19 March 2003 to the end of major combat operations on 1 May, I MEF user activity levels totaled 2.5 million tactical telephone calls, more than 240 video and audio teleconferences, more than 700 sensitive video teleconferences, and countless secure and unsecure e-mail transmissions. Most important, the command staff held daily teleconferences with deployed and rear area support staffs in the United States.[9]

In spite of the austere surroundings, MajGen James F. Amos holds a videoconference with his wing staff as they plan for future missions.

Photo courtesy of Col Charles J. Quilter II

and, of course, the younger Marines were speculating.[10]

Eyes in the Sky

Aerial reconnaissance missions over Iraqi airspace were common during Operation Southern Watch, and they intensified during the days leading up to OIF with aircrews searching for activity, particularly near oil pipelines and the northern Kuwaiti border. The steady collection of aerial imagery became an integral part of the air-ground strategy as commanders relied on the data to make tactical decisions. Unmanned aircraft, along with F/A-18 and AV-8B crews, gathered digital images of enemy movement and terrain features; the raw data was forwarded to intelligence analysts who, in turn, used their findings for time-sensitive squadron mission briefings. Collectively, the aircrews and unit intelligence sections worked closely with shared mission objectives, analyzing the data to minimize risks of troops-in-contact situations.

Although reconnaissance aircraft were not new airframes, upgraded equipment configurations with sophisticated system hardware and software weapon systems—along with a core of well-trained and experienced Navy and Marine Corps pilots and crews—enhanced 3d MAW's tactical capabilities. The Marines relied on four aircraft types to fly aerial reconnaissance missions: the AV-8B Harrier, the F/A-18D Hornet, the RQ-2B Pioneer UAV, and the Navy P-3C Orion. When authorized by Coalition forces air component commander General Moseley, the EA-6B Prowler was the fifth aircraft for such missions. However, Prowlers received more scrutiny at joint commands because of the limited fleet in service and the aircraft's status as a national asset.* Unfortunately, the joint approval process was not as responsive to the ground commander's immediate needs and air support requests, and it could take hours until a Prowler arrived on station.[11]

Despite harsh desert conditions, the Harrier fleet flew primarily armed reconnaissance and close air support missions. Recently upgraded with the Litening II targeting pod, the Harrier's overall aerial intelligence was enhanced by capturing video imagery during missions that painted a more comprehensive picture of current enemy disposition and future target recommendations. During combat actions, the pod was linked with precision-guided munitions, including 500- and 1,000-pound laser-guided bombs as well as laser-guided AGM-65C/E Maverick missiles. Harriers achieved a 65-percent effective rate, with no bomb landing farther than 200 feet from its objective (defined as a 13-foot circle centered on the target).[12]

The F/A-18 Hornet community repeatedly stress tested the aircraft's advanced tactical air reconnaissance system (ATARS), which was a real-time imaging system for image acquisition, data storage, and data link. With integrated infrared and light sensors linked to digital recorders and a radar interface that recorded imagery, this sophisticated system was internally mounted in place of the nose gun and included an external centerline datalink pod. Similar to the collection capabilities of the long-retired RF-4B Phantom II, the Hornet's enhanced reconnaissance capability delivered aerial

Photo by LCpl Richard W. Court

An unmanned aerial vehicle from VMU-2 is ready for launch from a pneumatic launcher in Kuwait. Although these unmanned aircraft relied on decade-old technology, the Marine Corps integrated its imagery collection for direct support of ground commanders during combat operations.

snapshots of enemy activity and troop movements, thus becoming a highly requested Marine air asset. To accommodate the demand, Hornet squadrons formed ATARS mission planning cells and assigned staff noncommissioned officers to fill wartime billets. For example, Staff Sergeant Andrew T. Millet was one such staff noncommissioned officer for Marine Fighter Attack Squadron All-Weather 533 (VMFA[AW]-533) who managed the daily influx of data.[13] The squadron created positions specifically to manage a high volume of reconnaissance requests from the joint force air component commander, the I MEF command element, the 1st MarDiv, the Royal Marine 3 Commando Brigade, and the U.S. Air Force 110th Fighter Wing.

The Marine Corps deployed its only two UAV squadrons—VMU-1 from Twentynine Palms, California, and VMU-2 from Cherry Point, North Carolina—and operated them under the 3d MAW umbrella. Both units arrived in mid-February and initially set up launch equipment at Tactical Assembly Area Coyote in Kuwait, which was nothing more than a stark dirt runway. By 26 February, both units had completed their required integration flights and began flying daily reconnaissance missions in support of Operation Southern Watch. Leading up to OIF, VMU-1 flew 128 sorties over 498 hours from Camp Coyote.[14] Dependence on the units' RQ-2B Pioneer aircraft increased steadi-

*Although the Air Force's Coalition forces air component commander managed and assigned Prowlers, they could be requested for specific missions on the air tasking order.

ly as both squadrons helped shape the battlespace with prestrike intelligence for ground commanders.

The U.S. Navy Lockheed P-3 Orion patrol aircraft was a welcome addition to the air-ground team, enhancing Marine Corps surveillance support options by delivering terrain imagery for commanders to view a continuous snapshot of the battlefield. Configured with an antisurface warfare improvement program system upgrade, P-3s carried sensors that provided another source of battlespace situational awareness. The platform was used extensively during Task Force 58 operations in Afghanistan. Staff members who transitioned from General Mattis's task force team in OEF to OIF leveraged lessons learned from that rugged theater of operations. By flying with the Orions during their assigned naval missions, 1st MarDiv "riders" actively participated in aerial observation and identified targets of interest inside Iraq. Often, several division riders were waiting to claim an extra seat on board. Initially, Combined Task Force 57 arranged for division-rider pickup points at al-Jaber and Ali al-Salem in Kuwait. But the joint rider program was so successful that the task force forward based a large Orion detachment at Ali al-Salem. By operating closer to the Marines, the P-3 squadron provided extended loiter time and surveillance coverage day and night.[15] After the al-Faw Peninsula was secured, P-3s provided direct support to I MEF for the duration of combat operations.[16]

Combined Task Force 57 under Cdr Robert Lally, USN, and the 1st MarDiv "rider" team coordinated by Col James W. Lukeman effectively employed Navy P-3 Orions such as this one as an observation, surveillance, and collection platform before and during combat operations. News of numerous P-3 mission successes spread quickly, resulting in an increased demand for the aircraft by Coalition and joint units.

Photo courtesy of 1st Marine Division

Intelligence Gathering and Reporting

Despite a demanding deployment schedule, the 3d MAW assistant chief of staff for intelligence, Colonel Eric M. Walters, assigned intelligence personnel to support multiple operations, including Operation Southern Watch, the Combined Joint Task Force Consequence Management, Combined Joint Task Force 180 in Afghanistan, and OIF. His staff also provided augmentation directly to Marine aircraft groups and squadrons and a contingent of intelligence Marines with General Amos's advance party to Ahmed al-Jaber Air Base to assist in setting up the newly constructed tactical air command center.

Intelligence personnel in the tactical air command center analyzed visual aerial reconnaissance data obtained by tactical aircraft and from strategic collection platforms. The section developed hard copy and electronic target folders for the air tasking order as well as proposed targets. Every 12 hours, new secondary target lists—about 30 throughout combat operations—were produced that indicated mobile and fixed objectives for close air support missions.[17] The analysis and plans sections prepared for an upcoming surge of situational awareness assessments and the need to publish data in a number of formats, including intelligence reports; summaries (every 12 hours); daily commanding general briefings via e-mail and on the Internet (every 24 hours); bomb damage assessment spreadsheets and graphics; situation map postings; enemy disposition and air defense graphics; electronic order of battle and air defense graphics; surface-to-surface missile strike logs; mission reports; in-flight reports; and specific tailored support, such as staff target folders and related material. Slated to track the visual aerial reconnaissance program, the collections section anticipated heavy demand for information requests and the need for rapid retrieval and dissemination of imagery for the tactical air command center battle staff and the future operations section that used the information to project a 48- to 72-hour outlook.[18]

Photo courtesy of Marine Light Attack Helicopter Squadron 269

Oliver L. North (left), a retired Marine Corps lieutenant colonel, traveled with air and ground Marines throughout Operation Iraqi Freedom, reporting as an embedded newscaster.

Public Affairs and Embedded Media

Unlike earlier military conflicts, the Department of Defense welcomed and supported a formal embedded civilian media program during OIF. This allowed reporters inside military operations to depict their efforts on a daily basis during the deployment. Major Jeffrey J. Nyhart, I MEF public affairs officer, implemented a program that verified journalist identification, approved credentials for theater access, and embedded them with units best tailored for such variables as their medium, the location and size of audience, expected security level of unit operations, and potential maximum exposure of the Marine Corps story. By June 2003, more than 340 media members were embedded throughout the I MEF area of operations.[19]

The Marine Corps Air Station Miramar consolidated public affairs office deployed 11 Marines who were assigned among the seven wing groups and 3 who were sent directly to CentCom staff. Staff preparations included mass unit briefings for those receiving embedded media representatives and detailed training for unit information officers because advanced satellite technology allowed real-time reporting. Anticipating a prominent role for helicopters during this air-centric campaign, a majority of 3d MAW's media representatives were sent to MAG-39 for embedment with rotary-wing squadrons at Ali al-Salem Air Base. The logic was that media representatives would have a greater chance to fly in helicopters and participate with Marines throughout the course of the battle. Six reporters were also embedded with MAG-11 units at Ahmed al-Jaber Air Base. MAG-16 received no embedded journalists because of the logistics of shipborne operations, but it did receive media attention.[20] Subsequently, the Miramar public affairs office produced and posted more than 40 stories to the official Marine Corps website as well as more than 100 combat-oriented photographs. Images and articles were printed in civilian publications from the high-quality digital collection.

Transitioning from Operation Southern Watch

By enforcing UN-dictated no-fly zones for more than a decade, Coalition aircrews gained significant familiarization with the environment that became their battlefield over Iraq. On 6 March 2003, VMA-311 flew its first mission inside the country, supporting Operation Southern Watch by launching AV-8B Harriers from the *Bonhomme Richard*.

Majors Marcus B. Annibale and Peter S. Blake flew to the coordinates of a kill box, looking for reference points that were assigned by intelligence Marines. Although the aircrew dropped no bombs, they reported enemy defensive positions west of Basrah, near the Rumaylah oil fields. They collected invaluable imagery with the Harrier's Litening II pod, identifying frontline Iraqi artillery positions and fuel storage tanks. Shortly thereafter, Major John H. Cane and a Royal Air Force exchange pilot collected video of personnel movement around an enemy fuel tank farm.

Likewise, Colonel Alles, commanding officer of MAG-11, recognized the value that the precombat flights provided for the group's Hornet aircrews: "Operation Southern Watch provided a great benefit—I can't think of any earlier conflicts in which tactical aircraft operated and bombed the targets before the actual war began. It was pretty much unprecedented."[21]

On 18 March, aircraft supporting Southern Watch conducted air strikes against Iraqi early warning radars and command-and-control sites in advance of the start of the ground war.[22] It was a prelude for the next day when 3d MAW aircrews—already quite familiar flying in the airspace under joint air control protocols—transitioned from enforcing the no-fly zone to advancing with 1st MarDiv.

Central Command's Order of Battle

At 0800 on 17 March, Marine Corps CentCom outlined the I MEF order of battle during a staff briefing,

Col Randolph D. Alles, MAG-11's commanding officer, often used a satellite telephone to communicate when feasible. This was one of several advanced telecommunications capabilities utilized by 3d MAW despite the austere desert conditions.

Photo courtesy of Col Charles J. Quilter II

which indicated offensive operations were hours away for 3d MAW.[23] General Amos empowered his Marines to communicate, decide, lead, and execute the air campaign during OIF. Just before hostilities began, the air wing commander expressed his thoughts more formally to the air groups in an e-mail attachment. He noted the honorable cause that they were embarking upon and declared that there was "a fear worse than death. . . . That is the fear of letting down your fellow Marines."[24]

Marine Corps Order of Battle

By 0800 on 17 March 2003, the order of battle for I MEF and the individual components strength and missions were depicted in briefing charts at Marine Corps CentCom.

- I MEF Command Element: 4,638
- 1st Marine Division: 20,606. Secure the southern oil fields; conduct a passage of lines through Task Force Tarawa and attack toward Baghdad.
- 3d Marine Aircraft Wing: 14,381. Shape I MEF's battlespace; screen the ground combat element from attacks; support Combined Force Air Component Command (CFACC).
- 1st Force Service Support Group: 10,504. Provide direct combat service support to I MEF; interface with the Marine Logistics Command, a theater-level command under operational control of CentCom.
- I MEF Engineer Group: 3,121. Maintain roads and bridges along the I MEF lines of communication. This unit was a composite of U.S. Navy construction battalions and Marine engineers.
- Task Force Tarawa (2d Marine Expeditionary Brigade): 5,091. Secure an-Nasiriyah and crossings across the Euphrates River; secure lines of communication.
- 15th Marine Expeditionary Unit: 1,739. Attach to 1 (UK) Armoured Division for Opening Gambit; attach to Task Force Tarawa.
- 1 (UK) Armoured Division: 21,045. Attack north from Kuwait; conduct relief in place in oil fields with 1st Marine Division; secure Basrah and vicinity.
- I MEF total: 81,125

Other Marine forces in theater:

Marine Central Command Element (Bahrain): 385
Marine Logistics Command (Kuwait): 4,525
Marine Central Command total: 86,777*

*This is the rendition of the CentCom morning report, 17 March 2003, captured by LtCol Jeffrey Acosta, the field historian attached to CentCom, and sent to the author by e-mail. The total does not show the Marines committed to Combined Joint Task Force Horn of Africa.

Chapter 7
Countdown to the Border

Delivering an Ultimatum

On 17 March 2003, Saint Patrick's Day, President Bush addressed the American people on national television and summarized a 12-year chronology of escalating tensions between the U.S. government and Saddam Hussein's regime. He highlighted years of frustration since the end of the Gulf War, a time characterized by incomplete weapons inspections and the ultimate failure of UN diplomacy to disarm Iraq's alleged weapons of mass destruction. Presenting a more aggressive course of action, the commander in chief outlined specific objectives that required Saddam's compliance within 48 hours. If the dictator failed to meet the deadline, the Coalition would authorize military force to liberate the Iraqi people, eliminate weapons of mass destruction, and end Saddam's regime.[1]

Midway through his 14-minute speech, President Bush conveyed a clear message that emphasized the necessity of Saddam Hussein's compliance:

> We will tear down the apparatus of terror and we will help you [the Iraqi people] to build a new Iraq that is prosperous and free. In a free Iraq, there will be no more wars of aggression against your neighbors, no more poison factories, and no more executions of dissidents, no more torture chambers and rape rooms. The tyrant will soon be gone. The day of your liberation is near.[2]

Photo by SSgt Michelle Michaud, USAF. Defense Imagery 030317-F-PJ904-014

On 16 March 2003, President George W. Bush held a one-day emergency summit meeting at Lajes Field, Azores, to discuss the possibilities of war in Iraq.

As the 48-hour deadline approached, Marines waited for the decision to launch OIF, which would implement more than 12 months of intensive planning that began at Marine Corps CentCom in Tampa, Florida.[3] As the president spoke from the Oval Office, ground forces prepped in staging areas near the northern Kuwaiti border to cross the line of departure, and 3d MAW continued flying Operation Southern Watch missions. During the next two days, aircrews, crew chiefs, and air controllers digested intelligence briefs, studied maps, and reviewed procedures at bases in Kuwait, Bahrain, and Saudi Arabia and on board U.S. Navy ships in the Arabian and Red Seas. Activity among aviation logisticians and maintenance crews peaked as they readied aircraft and conducted preflight checks to ensure the highest possible mission capable readiness rates. At this time, the Marine aviation combat element had grown to almost 15,000 personnel and 435 aircraft.[4]

General Amos addressed 3d MAW at Ahmed al-Jaber Air Base, Kuwait, knowing it would probably be the last gathering before hostilities began.

Photo courtesy of Col Charles J. Quilter II

An AV-8B Harrier from VMA-211 takes off from the flight deck of a "Harrier carrier" in support of a strike mission in Iraq.

He declared, "If ever there was a time to take the aviation assets of the 3d Marine Aircraft Wing to war, this is it right now. I feel very good about the systems. I feel very good about the availability of aircraft. I feel very good about the way we've trained our aircrews to fly these missions."[5]

Anticipating a two- to four-day separation before the start of ground operations, the air wing planned to conduct aggressive shaping fires well ahead of the land forces. The initial targets were observation posts across the border, the Iraqi *51st Mechanized Division*, artillery, and communication sites positioned in southern Iraq between Basrah and Kuwait.[6] The intent was to launch attack aircraft and quickly destroy known and suspected enemy threats during the opening hours of the campaign, building momentum as 1st MarDiv and the U.S. Army's 3d Infantry Division crossed the border from Kuwait.* As promised, the wing positioned its direct support aircraft with ground units so that commanders would have aviation options readily available. On 18 March, both of 3d MAW's UAV squadrons, VMU-1 and VMU-2, were in direct support of each 1st MarDiv regimental combat team and Task Force Tarawa. The squadrons provided full-motion video through a direct downlink capability, which enabled pinpoint awareness of enemy positions and the identification of targets in both the close and deep fight.[7] During the first 36 hours, each regiment had a squadron of F/A-18D Hornets assigned as well, so 1st MarDiv could directly benefit from the aircraft's advanced aerial reconnaissance system with a timely collection of tactical imagery. In turn, the wing added a placeholder on its air tasking order that earmarked direct support missions for a minimum of five daily aerial photo reconnaissance missions. Image data was extracted from tapes at al-Jaber and transmitted to intelligence analysts at the imagery interpretation platoon located at Camp Commando in Kuwait.[8]

*Uncertainty regarding locations of minefields near the Kuwait-Iraq border warranted collaboration between I MEF combat engineers and Navy Seabees with heavy reliance on the Kuwaitis to determine the number of breach lanes needed to support the movement. This enabled locals with a vested interest to guide the breaching effort. Gen Mattis's main command headquarters moved ahead of the division into northern Kuwait, ready to assume forward control of battlefield operations.

Opening Gambit

Under the command of General Amos, OIF officially began on 19 March for the Marines and

Photo by GySgt Matthew M. Smith. Defense Imagery 030319-M-RY230-001

A CH-46 Sea Knight supporting RCT-5 turns its rotors for a function check on the opening day of the air campaign. Dedicated helicopters were greatly appreciated by ground commanders in direct support of regimental combat teams.

Photo by LCpl Jonathan T. Spencer. Defense Imagery 030326-M-FW387-011

LCpl Christopher Kraveic of MWSS-272 fuels an F/A-18 Hornet from an M970 refueling truck at 3d MAW's fixed-wing base at Ahmed al-Jaber, Kuwait.

sailors assigned to 3d MAW.[9] In the coordinated air campaign, fighter and attack helicopter crews flew strikes against enemy artillery and communication sites in advance of 1st MarDiv's ground offensive. At the same time, transport helicopters prepared to ferry troops for a series of aerial insertions slated to begin two days later. During the weeks leading up to OIF, U.S. aviators primarily flew armed reconnaissance missions and targeted air defense and command-and-control sites. Air operations, however, expanded rapidly with close air support and assault support sorties once the regimental combat team crossed the border.

Named by General Amos during a planning session months earlier, the "Opening Gambit" signified the initial 96 hours in which 3d MAW

Official U.S. Marine Corps photo

The Marine Corps' tactical air command center was a constant flurry of activity as Gens Amos and Robling and their staff oversaw the air war. Outfitted with advanced telecommunications and computer interfaces, the staff also depended on maps and pushpins for visual displays of the air-ground movement to assist with command decision making.

launched a series of escalating intensive combined arms attacks to set the stage for the 23-day march to Baghdad.* During this period, the air-ground team focused on four key objectives: destruction of enemy observation posts and the Iraqi *51st Mechanized Division* near Safwan Hill; seizure of the Rumaylah oil field complex; completion of a relief-in-place with the 1 (UK) Armoured Division near the town of Basrah; and advancement of 1st MarDiv north of the Euphrates River.[10] Once accomplished, I MEF would be well positioned to continue its rapid march to Baghdad with the air wing overhead.

With the 3d MAW fleet in place, aircrews flew from three air bases and from Navy ships to begin the air campaign, referred to as A-Day, or "Air" Day. Four squadrons of AV-8B Harriers began launching aircraft from the flight decks afloat in the northern Persian Gulf as a detachment of Harriers took off from al-Jaber Air Base. A detachment of Hercules transport/refuelers at FOB Joe Foss was staged along with Hornets. Helicopters positioned at Ali al-Salem Air Base in Kuwait were on standby alert along with UAVs in position to collect targeting data for ground commanders. The Prowlers from Marine Tactical Electronic Warfare Squadrons 1 and 2 (VMAQ 1 and VMAQ-2) continued flying from Prince Sultan Air Base, Saudi Arabia, targeting radar and air defense communications networks, helping create a "fog of war" for the Iraqis.[11]

A Hub of Activity in the Tactical Air Command Center

Colonel McFarland, commanding officer of MACG-38, located the group headquarters at Ahmed al-Jaber Air Base alongside the 1st MarDiv combat operations center. It was an ideal location for immediate air-ground team coordination, but it was also in close proximity to 3d MAW headquarters and the tactical air command center, which encouraged operational as well as tactical collaboration.[12] Configured with advanced system capabili-

*The 23-day calculation was derived from the start of air wing operations on 19 March to the seizure of Baghdad on 10 April 2003, as cited in the 3d MAW command chronology, January to June 2003.

ties set up and maintained by Marine Wing Communications Squadron 38 (MWCS-38), the tactical air command center was the heart of operations for executing the air campaign. Staffed with four battle captains, two 12-hour shifts of watch officers, tactical air controllers, and operators, the center was 3d MAW's senior air control agency where General Amos and General Robling, the assistant wing commander, maintained a 24-hour visual picture of the air war. From this vantage point, the two generals assessed current and future air operations.

Anticipating a robust flight schedule, General Amos relied on four battle captains to balance the workload among air controllers in the TACC and to draft the daily air tasking order with a Marine-specific subset for direct support missions. By rotating shifts, a battle captain would write a draft air tasking order, then stand duty as the senior Marine in the command center during the next shift. This rotation enabled each battle captain to gain first-hand familiarity with and maintain accountability of the flight schedule before and during its execution. Using battle captains proved invaluable and allowed the general to focus on future operations, while the assistant wing commander oversaw the air war as it unfolded. General Robling declared, "There is no doubt that one of the primary reasons the campaign went so well as it did was because of the hard work of these four battle captains."[13]

The scenario inside the TACC changed for Colonel Jeffrey A. White and his fellow battle captains Colonels Raymond C. Fox, Mark D. Mahaffey, and William W. Griffen Jr. when the ground campaign launched earlier than planned and poor flying conditions dominated the battlespace. Instead of overseeing aggressive battlefield shaping missions as envisioned, watchstanders described the air war as reactionary. As Colonel White recalled:

> We initially briefed the tactical air command center crew on Thursday that we could go to

MajGen James F. Amos (left) selected Col Jeffrey A. White (right) to serve as one of four battle captains because he possessed the commander's criteria of the right skills, expertise, and command time needed to represent 3d MAW in the tactical air command center.

Photo courtesy of Col Charles J. Quilter II

Illustration courtesy of 1st Marine Division

Safwan Hill was the dominant terrain feature near the Iraq-Kuwait border, providing the enemy with an advantageous line-of-sight view 18 miles into Kuwait. The division commander relied on 3d MAW's strike aircraft to destroy the threat represented by the well-positioned Iraqi forces and to provide safe passage into Iraq during the Opening Gambit.

> war on Saturday morning [21 March] at 0300 Zulu, or 0600 local; however, during the shift, the situation changed rapidly and we learned that we're going to war tonight [19 March]. We noticed enemy activity moving down into southern Kuwait, and ground units on the border were exchanging artillery fire. We had a standby air tasking order for day one of the war that was embedded at the bottom of the schedule in case we needed to overlay it if air operations quickly accelerated without much warning. When the timeline changed, tremendous effort by the current and the future operations planners focused on realigning all the groups and preparing for on-call missions.[14]

By day two, the weather cleared and the air tasking order included more missions in which 3d MAW aircraft were directed to destroy enemy artillery and tanks. Initially, a 36-hour delay in receiving bomb damage assessment reports hampered future target selections and priorities, so Colonel White pushed the intelligence section to aggressively gather information to validate the air guidance matrix that outlined targeting priorities. He was pleased by the results a few days after the start of hostilities:

> Although we started on the enemy's timeline, we were backlogged; we were not getting damage assessment report[s] for a few days. Due to a lot of good people doing a lot of good work, we produced damage assessments more rapidly. The reports provided critical feedback regarding which enemy units or targets were damaged so that planners could identify key areas of targets that needed more attention.[15]

Precision Bombing at Safwan Hill

Taking the high ground at Safwan Hill was one of the first planned attacks of OIF. The mission, which had been rehearsed for weeks, entailed seizing an Iraqi observation post on top of the 551-foot hill that overlooked the Kuwait border. The post was thought to contain sophisticated surveillance monitoring equipment that was oriented toward a main highway that ran north from Kuwait to Basrah. The vantage point offered the Iraqis an unobstructed view of Coalition forces that would soon cross the border, just a few miles away.

Colonel Milstead, commanding officer of MAG-29, orchestrated a composite helicopter mission designed to insert a reconnaissance platoon on top of Safwan Hill. The aircraft came from HMLA-269 and HMM-162. Milstead's plan tasked 10 AH-1W Cobra attack helicopters armed with antiarmor missiles to lead the formation; 6 CH-46 Sea Knight helicopters would follow to ferry the reconnaissance Marines to the hill. Five command-and-control UH-1N Huey helicopters would fly in the rear, outfitted with GAU-17 Gatling guns capable of firing 3,000 rounds per minute.[16] HMH-464 was on standby to provide logistical support. Once successfully landed, the Marines would destroy any enemies who had survived the artillery and air strikes conducted before the assault.[17]

In the late evening hours of 19 March, the 11th Marines unleashed an artillery barrage in preparation for the coordinated aerial attack that soon followed. During the next 24 hours, MAG-29 Cobras swept in low from the south to strike the position with rockets. Navy and MAG-11 F/A-18 Hornets provided extra firepower and dropped more than 40,000 pounds of explosives on this key position.[18] VMFA(AW)-121, nicknamed the "Green Knights," led the first MAG-11 air strike of OIF. Commanded by Lieutenant Colonel Matthew Shihadeh, the Green Knights' ordnance configuration typically consisted of a laser-guided bomb, GBU-31 joint direct attack munitions, and MK83 general-purpose bombs. This weapon mix provided a number of options for ground commanders for close air support, deep strikes, forward air controller (airborne) reconnaissance and strike coordination, and armed reconnaissance.[19]

Poor flying conditions, dense fog, and smoke from burning oil fields delayed the troop insertion one day. By dawn, MAG-29 successfully landed

MajGen James F. Amos (right), 3d MAW's commanding general, meets with Col Robert E. Milstead Jr., commanding officer of MAG-29, to discuss the air strategy. Gen Amos often flew to speak face-to-face with his commanders, in addition to using videoconferences, secure telephone, and encrypted message communications.

Official U.S. Marine Corps photo

MAG-29's Safwan Hill Raid

When MAG-29 executive officer Lieutenant Colonel Darrell L. Thacker Jr. heard about the unsuccessful joint decapitation strike aimed at Saddam Hussein, he was not surprised when General Conway sent a radio message to MAG-29 headquarters asking if the aerial insertion mission could launch in 30 minutes. After the failed strike, senior military leaders were even more concerned about the threat of Iraqi oil well sabotage, which could jeopardize any hope of future economic stability after hostilities ended. Colonel Thacker fully understood the need for urgency and recalculated the mission. The composite squadron aircrews were airborne and heading toward Safwan Hill 90 minutes after the revised plan was issued.[20]

Serving as the air mission commander for the raid, Colonel Thacker launched the group's helicopters at sunset into the eye of a nasty storm. Cobras and all trace helicopters blazed through the swirling sand and checked in at the rendezvous point as planned. The aircrews waited more than 20 minutes for the F/A-18Ds to arrive on-station. The delay was worthwhile because the Hornets conducted a series of air strikes, attacking the hill with joint direct attack munitions.

The Cobras trailed close behind and fired their rockets at the top of the hill, but the pilots of the CH-46s ferrying the ground troops struggled to gain visibility in the landing area. They tried three times to land in the drop zone, but the pilots were blinded by the blowing sand. After the third attempt, the weather deteriorated well below acceptable limits for safe flight and landing, and several pilots reported experiencing severe vertigo and disorientation. At that point, Thacker aborted the mission.

The return trip to FOB Joe Foss proved even more challenging as the weather worsened, which further confirmed the decision to call off the mission.[21] As a seasoned pilot with more than two decades of experience, the colonel directed the aviators to spread out so that all aircraft would retain sufficient separation as they ventured deeper into the murky sky. After the aircraft landed safely, Thacker initiated plans to reschedule the mission for early the next morning after briefing his recommendations to the wing leadership.

On the first evening of Opening Gambit, 3d MAW aircrews experienced turbulent weather, moonless skies, and blowing sand, but it was just a prelude to the weather-related disruptions that would plague flight operations on the move toward Baghdad.[22] The colonel's decision to cancel the troop insertion mission at Safwan Hill balanced his desire for mission completion with a concern for the well-being and safety of aircrews and aircraft.[23]

Photo by Sgt Paul L. Anstine II. Defense Imagery 030321-M-5150A-056

M1A1 Abrams tanks from Company C, 1st Tank Battalion, wait in attack position south of their breach point as AH-1 Cobra gunships fly overhead, attacking Safwan Hill before ground units cross the Kuwait-Iraq border during the opening of Operation Iraqi Freedom.

two dozen reconnaissance Marines onto the hill.[24] Bomb damage assessment reports confirmed that Safwan Hill—and any enemy forces there—no longer posed a threat. The observation post was destroyed, which was essential for Coalition forces to cross the border unimpeded by Iraqi artillery and which prevented the Iraqis from calling for reinforcements. As a result, 1st MarDiv successfully crossed the border without interruption or diversion and began pushing toward Baghdad.

Joint Forces Authorize a "Decapitation Strike"

On the evening of 19 March, Coalition forces received information from multiple sources that pinpointed the location of a prominent enemy gathering. Intelligence data zeroed in on suspected enemy "targets of opportunity" who were attending a meeting at an Iraqi leadership compound known as the Dora Farms complex on the outskirts of Baghdad. The information generated high-level interest by U.S. government leaders and presented military commanders with an opportunity for an isolated, combined arms strike at Saddam and his leadership hierarchy.[25]

The guided missile cruiser USS Bunker Hill *(CG 52) launches the first Tomahawk missile into Iraq in support of Operation Iraqi Freedom.*

Photo by PO2 Richard Moore, USN. Defense Imagery 030320-N-4655M-009

Senior leaders debated the pros and cons of orchestrating a joint decapitation strike, applying two key doctrinal components of combat power—surprise and boldness.[26] If the strike was successful, large-scale hostilities and a major troop crossing into Iraq might be avoided. However, if the strike failed to hit the targets, the element of surprise would be lost, and Saddam's forces would gain a tactical advantage. They would realize that a subsequent air and ground attack was imminent and be able to position their forces for a counterattack against the Coalition. A major consideration was that the Kuwaiti government had turned over its local airspace to U.S. control earlier that day as part of an agreement derived from several planning sessions held weeks earlier.[27] Weighing this newfound opportunity against potential risks, the White House and CentCom leadership assessed the situation and ultimately authorized a coordinated cruise missile and air strike.

Given the seven-hour difference between Washington, DC, and Iraq, the joint strike force launched before sunrise on 20 March and included missile and bomb attacks. Forty U.S. Navy Tomahawk cruise missiles were launched from six warships: the USS *Milius* (DDG 69), USS *Donald Cook* (DDG 75), USS *Bunker Hill* (CG 52), USS *Cowpens* (CG 63), and the submarines USS *Cheyenne* (SSN 773) and USS *Montpelier* (SSN 765). The U.S. Air Force bombing missions were flown by Lockheed F-117A Nighthawk and McDonnell Douglas F-15E Strike Eagle stealth attack aircraft and escorted by four—three Marine and one Navy—EA-6B Prowlers.[28]

That day, Major Scott A. Cooper, a Marine naval flight officer, was flying one of those Prowlers near the Iraqi no-fly zone during the early morning hours. When Major Cooper's aircrew from VMAQ-1 received targeting instructions and a timeline from an air controller on the Air Force airborne warning and control system coordination frequency to "go to Baghdad and link up with the F-117s and support the Tomahawk strikes," they had no idea they would join a coordinated preci-

Photo by SSgt Matthew Hannen, USAF. Defense Imagery 030318-F-9528H-027

Marines from VMAQ-2 perform squadron-level maintenance late into the evening in the rear cockpit of an EA-6B Prowler at a forward deployed Southwest Asia location.

sion air strike aimed at the Dora Farms complex. Considered a high-interest mission, the Prowlers jammed radio frequencies while the Nighthawks dropped laser-guided weapons. Cooper described the events of the early morning mission in a later interview:

> We went to Baghdad and supported them. It was a great fireworks show. It was an interesting flight; the biggest thing we were fearful of was running out of gas. We weren't scheduled for gas and the tanker aircraft weren't flying into Iraqi airspace yet. We had to refuel before returning to Prince Sultan Air Base, which was about a one-and-a-half hour flight. The sun was just coming up, and we asked the controller about our tanker.[29]

Although Navy ships and Air Force crews struck their targets with precision missiles and bombs, early reports indicated that Saddam escaped but might have been injured during the surprise attack. Regardless, the strike rattled Iraqi forces, who sounded a wake-up call for Marines by counterattacking with a series of surface-to-surface missile launches. Marines reacted immediately by donning their nuclear, biological, and chemical suits and moving to designated bunkers, a drill that had been rehearsed for months and would be exercised repeatedly over the next several days. One of the incoming missiles landed within a kilometer of I MEF headquarters at Camp Commando in Kuwait.[30] Immediately after the explosion shook the ground and produced a grayish plume of smoke outside of the camp perimeter, 3d MAW launched four armed ground-alert AH-1W Cobras to scout the area.[31] Thanks to U.S. Army Patriot defense batteries assigned to MACG-38, nine Iraqi missiles were destroyed prior to impact, and no Coalition injuries were reported.[32]

These Iraqi missile attacks raised the possibility of biological or chemical agents. The U.S. military commands distinguished different levels of protection against biological or chemical attacks and categorized threat conditions by numeric values ranging from zero to four; a mission-oriented protective posture (MOPP) condition meant that all protective garments were to be worn, including overboots, masks with hood, and gloves. Although

Photo courtesy of Marine Light Attack Helicopter Squadron 269

Marines from 3d MAW crowd inside a bunker wearing protective gear as a precaution against the threat of chemical weapons during the first few days of the air campaign.

the bulky protective suits were cumbersome, they became an integral part of the daily routine during the early days of hostilities.[33]

The Air-Ground Campaign Takes Shape

Before OIF began, CentCom published its final draft of OPLAN 1003V (with revisions) for field distribution.[34] Not surprised by a steady flow of last-minute changes, I MEF plans shifted drastically mere hours before the start of offensive operations, thus overriding months of detailed planning. Originally, air-ground planners had planned for a two- to four-day separation between the initiation of the air campaign and follow-on ground combat operations. This window would allow 3d MAW time to conduct aggressive and intense preparatory fires well ahead of the ground forces. However, when the decapitation missile strike failed to hit Saddam and human intelligence information revealed enemy activity near the Kuwait-Iraq border, the timeline quickly unraveled. As a result, the opening offensive originally scheduled for 0300Z* on 21 March occurred 12 hours earlier than planned, altering the flow of forces.

Aerial imagery and recent intelligence data about the "ghost brigade"** and its suspected T-72 tanks near the Kuwait-Iraq border continued to concern the I MEF staff. Heightened enemy activity levels indicated potential sabotage of the oil fields by Iraqi military forces, a scenario that the Coalition had fully anticipated and planned to minimize.[35] The Rumaylah oil fields produced 1.6 million barrels per day and contributed to more than half of total Iraqi oil production.[36] Coalition planners feared that Iraqi forces would set fire to their own oil wells in a scene reminiscent of Kuwait during Operation Desert Storm in 1991, thus disrupting the nation's economy and ecology. The relevance of this threat was twofold. If oil produc-

*U.S. military forces use Zulu, often shortened to the letter "Z," to denote Greenwich Mean Time, which is the standard reference point when communicating across multiple time zones.

**Human intelligence reports were sketchy, so enemy activity pointed to a possible trap that was referred to as the "ghost brigade."

tion was severely disrupted or destroyed, it would most likely generate drastic, long-term financial and economic consequences for the Iraqi people who had already suffered years of oppression. It could also affect the reconstitution timeline after the successful conclusion of Coalition combat operations.

Based on Department of Defense and CentCom decisions, the I MEF commander, General Conway, modified the timeline just hours before its planned execution and reset the start to 1500Z on 20 March.[37] Although units scrambled to meet the revised launch time, the ability of I MEF to respond quickly even during the opening offensive reinforced the benefits of the planning process and unit readiness.

Unfortunately, the condensed timeline meant 3d MAW was unable to shape the battlefield for as long as they had planned, so instead, the air wing temporarily assumed a reactionary role to enemy activities. Nevertheless, the wing responded to the ground commander's request for aerial reconnaissance missions to view a snapshot of the battlefield in advance of ground force unit movements. Last-minute changes by their planners meant that 1st MarDiv would cross the line of departure 12 hours earlier than planned. The ripple effect of General Mattis's fragmentary order resulted in a revised night (beginning at 1800 local time) tactical movement of more than 20,000 I MEF Marines and sailors over the Kuwait border into southern Iraq to preset attack positions, rather than a dawn movement as originally planned.[38] The logic of a night troop movement from the line of departure in conjunction with an accelerated pace to Baghdad was consistent with generating tactical surprise, which in turn would destabilize enemy forces and minimize potential attacks from the Iraqi Army, Saddam's *Republican Guard*, and even renegade paramilitary forces.

Although riskier, the night operation also meant that RCT-5 would swap with RCT-7 to become the lead unit.[39] Planners quickly revamped tactical details of a plan that had taken months to fine tune.* RCT-5, commanded by Colonel Joseph F. Dunford Jr., led 1st MarDiv into combat with the firepower and support of 3d MAW overhead. Within 24 hours, I MEF had moved into southern Iraq and, most importantly, with no loss of life.[40]

Shaping the Battlefield

Lieutenant Colonel Lawrence R. Roberts, commanding officer of VMFA(AW)-533, became increasingly annoyed with the barrage of Iraqi Scud surface-to-surface missile alerts and repeated trips to the bunkers after the decapitation strike. During Operation Southern Watch, the squadron had flown 295 F/A-18D Hornet sorties, logged 616.3 flight hours, and expended 18,000 pounds of ordnance, so aircrews were accustomed to a heavy flight schedule.[41] With his aircrews useless while crammed inside bunkers at al-Jaber Air Base, Colonel Roberts wanted his Hornets to fly armed reconnaissance missions and operate as the division's forward-looking eyes before ground troops moved into southern Iraq. Guidance from the air wing staff was mixed regarding whether it was riskier to launch or leave the aircraft on the ground during the barrage of Scud attacks. He and his operations officer, Major John P. Farnam, walked to MAG-11 headquarters in hopes of gaining a tactical mission assignment.[42]

During their visit, 1st MarDiv's air officer, Lieutenant Colonel Bruce A. Shank, requested air support to engage approximately 75–80 Iraqi T-72 tanks digging in near RCT-7's border crossing point. An established Iraqi informant reported sighting an entire *Republican Guard* armor brigade and enemy movement near the Rumaylah oil fields just north of Safwan Hill.[43] The incoming information was conflicting, but worthy enough to warrant further investigation and a swift reaction if a possible trap was confirmed. Colonel Roberts was especially eager to respond to this air request because his squadron was assigned to fly in direct support of RCT-7.[44] Months before hostilities began, he had directed his aircrews to gain an understanding of the regiment's area of operations during face-to-face meetings. This further built on General Amos's philosophy of building relationships by making aviators better prepared to support the unit mission and instilling mutual trust between ground and air forces.

Colonel Roberts and Major Farnam wasted no time and launched with another Hornet waiting on the flight line in ready alert status. The Hornets' mission was to search for enemy movement along

*One of 1st MarDiv's logisticians, Maj Christopher B. Snyder, initiated a matrix-based execution checklist that identified, phased, and managed the breach-line flow of units over the border. A single-page document, it streamlined and graphically displayed the time, route, and flow of forces as they crossed the border into southern Iraq.

Photo by LCpl Andrew P. Roufs.
Defense Imagery 030320-M-9124R-097

Marines assigned to an explosive ordnance disposal unit with 1st LAR use a 7.62mm M14 sniper rifle to dispose of landmines along the Kuwait-Iraq border.

Highways 6 and 8, about 100 miles north near an-Nasiriyah. Roberts flew north toward the port of Umm Qasr, while Farnam, the weapons systems officer, monitored the radio and looked for targets. They had located nothing until nearly sunset, when they sighted a column of enemy armored vehicles parked on an overpass, perhaps to trap RCT-7. Major Farnam checked in with the direct air support center and then confirmed with the combat team's fire support coordination center that the tanks were enemy rather than friendly. The multi-tier command-and-control communication protocol provided an extra precaution to minimize the threat of friendly fire. As the colonel maintained altitude restrictions, the major locked onto the tank column on the forward-looking infrared radar in the rear cockpit. The Hornet made several attack runs, expending a mix of MK20 Rockeye II cluster bombs, 5-inch rockets, and 20mm cannon rounds. The cluster bomb canisters failed to open during the first run, but hit an enemy vehicle on the second. The rockets struck wide of the target, but the aircraft sprayed its cannon during the final two runs, depleting all the aircraft's ordnance and most of its fuel before returning to base. Although the aircrew was unsure of their success because they did not receive a bomb damage report, their flight manifested the squadron's eagerness to launch quickly and search ahead of the ground forces, thus setting the tone for future combat missions.[45] During the first 36 hours of the war, MAG-11 Hornet squadrons tallied 96.3 combat flight hours.[46]

Marine Corps air liaison officers, such as those pictured here, assisted Air Force E-3C Sentry aircrews and controllers in joint airspace management.

Photo courtesy of Col Charles J. Quilter II

Ground activities began on 21 March, one day after the air campaign ended, when U.S., British, and Coalition forces crossed the line of departure into southern Iraq earlier than planned.[47] By the time General Conway's expeditionary force took control of its battlespace, 3d MAW had already flown 259 sorties, quadrupling its flight schedule within a 24-hour period.[48]

As planned, the U.S. Air Force monitored Marine airspace through eight Boeing E-3C Sentry airborne warning and control system aircraft until ground troops crossed the line of departure. Because the Sentry would play the main tactical command-and-control role, especially at the start of the air campaign, MACG-38 assigned 10 liaison officers to work with Air Force aircrews to oversee interdiction targets and protect Marine aviation interests. The liaison officers ensured that planned and immediate close air support missions for the ground combat element were not reassigned to other Coalition units. In so doing, they demonstrated the best attributes of the Marine air command-and-control system for the joint Coalition and, most importantly from their point of view, the Marines on the ground.[49] As word spread among the joint aviation community that Marine air command and control was effectively managing its tactical airspace, aviators from other services began checking in for mission assignment with the TACC.[50]

Part III

Decisive Operations

Chapter 8
Air Operations Intensify

Movement to Contact

As the battle plan unfolded, the U.S. Army's V Corps led the main effort toward Baghdad with I MEF Marines simultaneously advancing on a different avenue of approach toward the capital city. The soldiers, commanded by U.S. Army Lieutenant General William S. Wallace, crossed the Euphrates River, then plowed northwest through the open Arabian Desert, while General Conway's force diverted enemy attention from the main effort by heading north and eastward. Meanwhile, Task Force Tarawa, commanded by General Natonski, crossed into Iraqi territory and quickly secured Jalibah airfield, a critical base for sustaining forward air operations. The task force's next objective was ensuring 1st MarDiv's safe passage over the Euphrates River near an-Nasiriyah. From there, I MEF would charge toward Baghdad to link with V Corps for a joint offensive to seize Baghdad. Planners estimated the trek from Kuwait to secure the Iraqi capital would take roughly 55 days. Yet, like most campaigns, plans changed rapidly, especially as the Marines neared the Euphrates River and an Nasiriyah.

During months of planning, the United States and United Kingdom strengthened their partnership, and the two countries eventually provided the majority of the forces for the Coalition effort. On the Marines' eastern flank, the 1 (UK) Armoured Division took control of the oil fields near Basrah and the port of Umm Qasr. Commanded by Major General Robin V. Brims, the British easily outgunned the enemy and seized both objectives.[1] Once secured, the division was strategically positioned to offer much-needed humanitarian relief for the area's residents.[2]

Destroying the Iraqi 51st Mechanized Division

On 20 March, RCT-5, under the command of Colonel Dunford, crossed the Kuwait-Iraq border nine-and-a-half hours earlier than planned.[3] The early attack was prompted by intelligence reports of enemy movement and repositioning of a *Republican Guard* armor brigade just north of Safwan Hill.[4] This information switched RCT-5 to the lead unit to block Highways 1 and 8, isolate the *51st Mechanized Division*, and secure the critical gas-oil separator plants. The az-Zubayr pumping station—or "the crown jewel"—was particularly important as it produced more than $40 million of oil per day and represented a dependable income source for Iraq after it was liberated.[5]

Photo by GySgt Matthew M. Smith.
Defense Imagery 030320-M-9106S-002

Col Joseph F. Dunford Jr. (on right), commanding officer of 5th Marines, speaks to members of the air wing attachments in preparation to cross the line of departure.

The early RCT-5 launch occurred simultaneously with the 1 (UK) Armoured Division's attack on Umm Qasr, which further reinforced the surprise element.[6] Despite the revised time line, the team viewed the change as an opportunity and assumed the role as the 1st MarDiv's main effort, supported by 3d MAW aircraft.

Aggressive preparation fires from a number of Cobra gunships, coupled with ordnance dropped by Harriers and Hornets, pounded the Iraqi *51st Mechanized Division*. Flying in darkness, 3d MAW aviators destroyed enemy positions as the Corps'

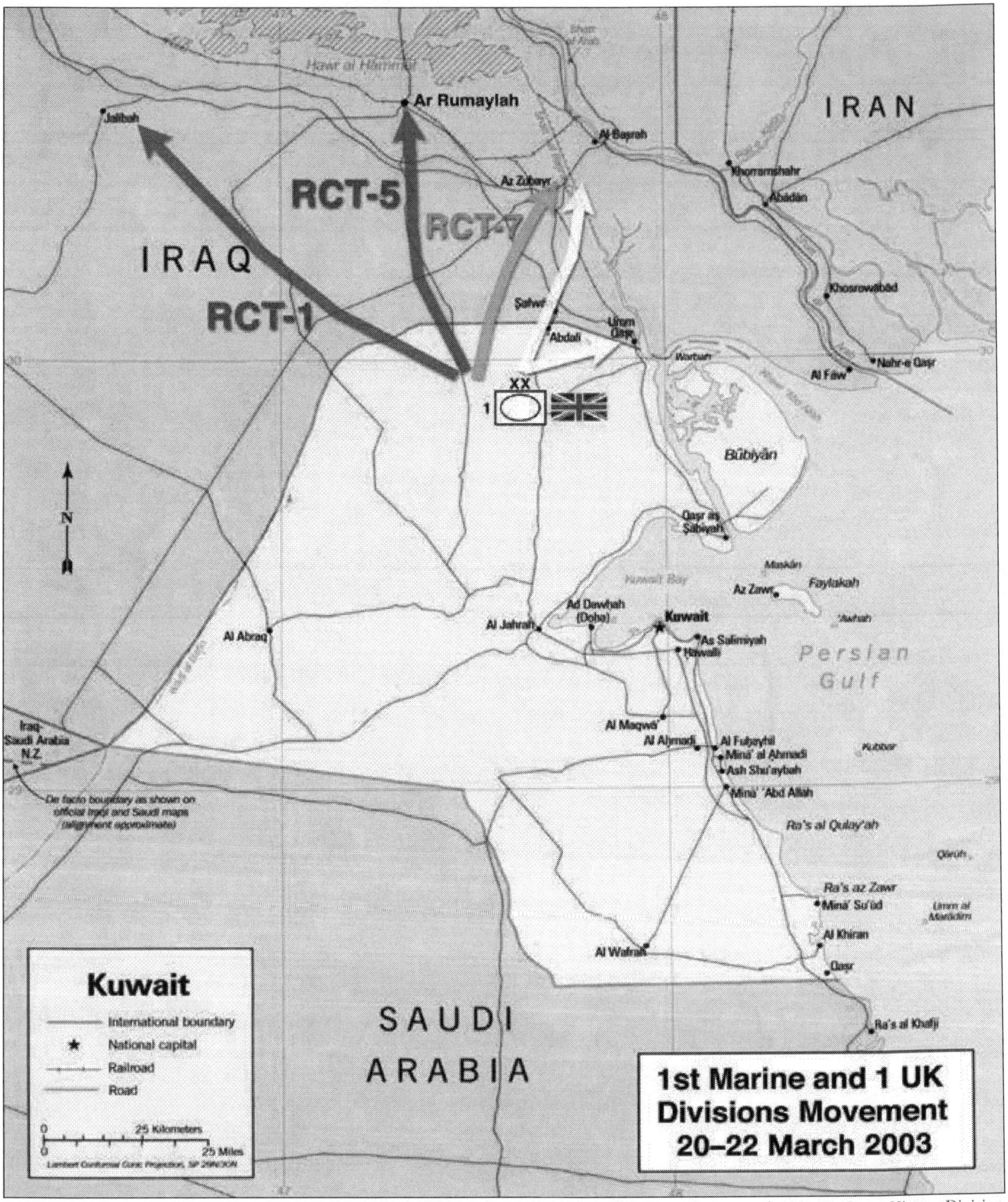

Adapted from a Central Intelligence Agency map by Marine Corps History Division

first ground unit moved inside enemy territory. At dusk, HMLA-267, commanded by Lieutenant Colonel Stephen K. Heywood, joined the air war. As part of MAG-39, the squadron launched five light divisions, each consisting of two AH-1W Cobras and one UH-1N Huey, with one Cobra flying backup. In successful attacks, aviators destroyed five key Iraqi border observation posts, enabling Coalition ground forces to approach the line of departure without enemy intervention or observation. The squadron's aircrews provided broad coverage for I MEF as the air war intensified, fly-

Defense Imagery 030327-M-2064V-005

A machine gunner assigned to HMLA-267 sits inside a UH-1N Huey helicopter and searches for enemy forces near Jalibah airfield.

ing missions in support of each regimental combat team, Task Force Tarawa, and British forces while also flying as airborne escorts for unit commanders.[7] Summarizing one of his first missions, Colonel Heywood recalled,

> My division was going to take out post five, so we had a practice run. I flew as a fourth aircraft with General Amos and showed him the target I was supposed to hit, which was close to the direct air support center. Obviously, he was concerned about the proximity to the air controllers, so I explained in detail how we planned to attack. The general left with a good feeling.[8]

VMFA(AW)-533, known as the "Hawks," was also instrumental during the opening days of the air war, conducting aerial reconnaissance before 1st MarDiv and the U.S. Army's 3d Infantry Division crossed the line of departure. Commanded by Lieutenant Colonel Lawrence R. Roberts, the F/A-18 Hornet squadron provided deep strike capabilities and strike coordination missions forward of friendly forces as the Hawks attacked Iraqi indirect fire assets, armor, and command-and-control targets. During the first 36 hours of hostilities, the squadron led MAG-11 efforts by flying 96.3 sortie hours, which included providing close air support for the initial British assault on the al-Faw Peninsula.

Adding fixed-wing firepower on the morning of 21 March, VMA-311's commanding officer, Colonel Michael K. Hile, flew the squadron's first mission of OIF. Teamed with wingman Captain Jason K. Duncan, their first strike was aimed at the Alamo Bridge, 30 nautical miles west of Basrah. Targeting data provided by Litening II target-

MALS-39 Marines Keep Helicopters Flying

MAG-39's Marine Aviation Logistics Squadron 39 (MALS-39), commanded by Lieutenant Colonel Paul F. Callan, supported four different types of helicopters—Hueys, Cobras, Sea Knights, and Sea and Super Stallions—which was quite a demanding task given the age of the aircraft and the harsh desert environment. Despite the extremely challenging weather conditions, MAG-39 maintained relatively high helicopter readiness rates. A few days after the air war started, HMLA-267 reported 79.7 percent full mission capable for its 9 UH-1N Hueys and 80.2 percent partial mission capable for its 18 AH-1W Cobras. This latter metric indicated that the squadron's aircraft could perform designated missions with noted discrepancies because secondary, back-up systems were operable. Such aircraft readiness was directly attributable to the 500 Marines who prepared the aircraft before deployment and managed maintenance and procurement activities during combat operations.*

The MALS-39 Marines kept its fleet aloft by repairing aircraft at maintenance facilities on shore at Ali al-Salem Air Base, Kuwait, and afloat on the SS *Curtiss* (T-AVB 4). For the first time in the history of Marine aviation, the MALS-39 aviation maintenance logistics ship was employed solely in a sea-based mode, supporting more than 130 helicopters with an average repair time of only two days.[9] Additionally, the squadron was the engine manager for all Marine helicopters operating in the CentCom area of operations, which entailed the buildup, delivery, and installation of 109 engines. Once the air campaign began, a detachment of about 70 MALS-39 maintenance Marines helped establish a forward base at an-Numaniyah inside Iraq, which further supported Marine aviation operations as a maneuver element.[10]

*A full mission-capable aircraft readiness rate indicated that all primary and secondary weapons systems were fully operable, whereas a partial mission-capable rate meant that secondary or backup weapons systems were operable to counter failed primary components.

ing pods enabled the pilots to score direct hits on the bridge with one GBU-16 and two GBU-12 precision bombs.[11] In the early morning hours of 21 March, Harriers from the 15th MEU joined the Opening Gambit. Major David C. Forrest and Captain Benjamin S. Krippendorf also flew one of the initial AV-8B missions and dropped bombs striking Iraqi artillery near Safwan Airfield.

Marine aviation partnered with lethal firepower from the 1st Light Armored Reconnaissance (1st LAR) Battalion and the 11th Marines. The combined arms capability provided extra protection for 1st MarDiv as it moved deeper into Iraq. RCT-5 blocked the Iraqi *6th Armored Division* so that RCT-7 could destroy elements of the *51st Mechanized Division*.[12] RCT-5 also encountered two Iraqi regular army brigades reinforced with artillery, yet it managed to secure critical gas and oil separation plants in a mere 16 hours.[13]

Advanced Lifelines

When RCT-5 crossed the line of departure, Colonel Michael C. Anderson, the commanding officer of MWSG-37, wasted no time extending the logistical network from Kuwait to support 3d MAW operations in Iraq. Joining efforts with Task Force Tarawa and MACG-38, the aviation combat element secured Jalibah airfield, the first FOB established in Iraq. Code named Riverfront, the airfield was located about 20 miles south of an-Nasiriyah on Highway 1. Although the field had been abandoned after the Gulf War, MWSG-37 quickly transitioned it into a hub of activity for aircrews near an-Nasiriyah.[14] The site not only supported aircraft operations, it facilitated a major logistical flow of equipment, cargo, and supplies before 1st MarDiv crossed the Euphrates River. It became a critical logistical lifeline for sustained Marine air operations by providing ordnance and fuel for hundreds of aircraft supporting ground forces throughout the rapid march to Baghdad and even during post-hostilities.[15]

Typically, MWSG-37 Marines searched for flat ground or existing concrete slabs on which to build the next site to support forward air operations. During the march to Baghdad, the group established 21 temporary FARPs at approximately 40- to 50-kilometer intervals, eight KC-130 tactical landing zones, and six FOBs.[16]

The Golden Spike

Lieutenant Colonel Roger R. Machut's 6th Engineer Support Battalion, 4th Field Service Support Group, assisted in building the extended fuel line from a 6-million-gallon U.S. Army bulk fuel farm at Breach Point West to Jalibah airfield, where Marines subsequently established their own bulk fuel farm. Colonel Machut designated this implementation as the "Golden Spike," in reference to the historic event at Promontory Summit, Utah, where the eastern and western sections of the first transcontinental railroad were linked in 1869.[17]

The Marine Corps' hose reel implementation team followed a three-step process. The team first conducted a preliminary reconnaissance of the route to find and dispose of unexploded ordnance. Then, they graded the surface and dug a V-shaped ditch, and finally the 6-inch-diameter rubber hose was pulled from truck-loaded spools and carefully laid into the ditch and covered. Once installed and operational, the hose reel alleviated dependence on trucks to transport fuel to forward-based units.

Chief Warrant Officer-4 Thomas M. Cierley, a bulk fueler who had served in Vietnam more than three decades earlier, summarized the installation: "Back then, we were using slower pumps, smaller fuel bladders, and relying on trucks for transport. During Operation Iraqi Freedom, we were bigger, better, and more efficient, plus it was a way of transporting fuel without motorized support."[18]

LCpl Christopher Kraveic, a refueler with MWSS-272, listens for the location of the next aircraft waiting for fuel at al-Jaber Air Base, Kuwait.

Photo by LCpl Jonathan T. Spencer. Defense Imagery 030326-M-2900S-009

A Marine Corps CH-53D Sea Knight drops bundled supplies at Jalibah, which was the first forward operating base established by MWSG-37 in Iraq.

Photo by Sgt Giles M. Isham. Defense Imagery 030429-M-33681-001

Assigning MACG-38 Marines to leapfrog with the air wing was just as important to maintaining FOBs and FARPs. Radio operators from MWCS-38 equipped with tactical satellite radios—the VHF/UHF-capable AN/PSC-5—were assigned to specific air bosses who managed onsite operations to ensure secure voice connectivity. Additionally, Marine air traffic control mobile teams were strategically positioned to oversee air control operations.

Captain Jeffrey D. Wrobel and Master Sergeant Forrest W. Frazier comprised the headquarters element, and they moved throughout the Iraqi battlefield, ensuring stable remote air-ground operations, troubleshooting problems, and providing invaluable feedback to MACG-38 and its forward controlling agencies. In all, the control group staffed seven teams to maneuver with MWSG-37 as it spanned from central Iraq to Tikrit. When the

Photo by LCpl Alicia M. Anderson.
Defense Imagery 030403-M-5607A-004

A Marine Corps Harrier prepares to land at one of several FARPs set up in Iraq during the march to Baghdad. The ability to quickly refuel and rearm at designated hubs allowed pilots to extend their time on target to support ground forces

Photo by LCpl Alicia M. Anderson.
Defense Imagery 030403-M-5607A-012

A Marine Corps Harrier is refueled "hot"—with its engine running—by LCpl Brent A. Starns from MWSS-271 at one of many FARPs operating in Iraq.

support group's FOBs and FARPs closed, moved, and relocated to other sites, MACG-38 coordinated its relocation and traveled alongside, further demonstrating that the air wing could operate as a separate aviation maneuver element.

Protecting the Oil Fields

As the remainder of 1st MarDiv crossed into Iraq and pushed inland, elements of 3d MAW, the 15th MEU, and the Royal Marines consolidated efforts and headed east. Their objective was to secure the oil fields near the seaport of Umm Qasr, about 30 miles south of Basrah. The port sparked keen economic and strategic interest because it provided the main access point to the Persian Gulf and was essential for commercial shipping of oil and other commodities.

During an aerial reconnaissance mission on 20 March, VMU-2, commanded by Lieutenant Colonel Robert D. Rice, flew one of its Pioneer aircraft over a small town and confirmed enemy activity through a live data feed to the division command post. Intelligence analysts and Pioneer operators discovered a large equipment storage and repair facility that housed dozens of T-55 tanks and self-propelled artillery pieces. Together, they shifted fires and destroyed the enemy cache aligned in defensive positions along RCT-7's intended attack routes, which further endorsed assigning aircraft in direct support of 1st MarDiv's units.[19] Eliminating an unnecessary "bureaucracy" of command layers allowed operators, intelligence sections, and unit commanders to aggressively attack and respond to dynamic events on the battlefield.[20]

The Coalition also planned to use Umm Qasr as a major offload point for humanitarian supplies. Although engaged in combat, Coalition forces kept in mind the need to safeguard the Iraqis' hope for

Photo by LCpl Matthew J. Decker. Defense Imagery 030324-M-CB246-001

Marines assigned to the 15th MEU fire a 155mm howitzer at Umm Qasr, Iraq, five days after the air campaign began.

economic stability and future growth after the war ended. The Coalition quickly seized the oil fields and established military control to guard them against sabotage. After the port was secured, they distributed much-needed humanitarian assistance, a key element for the Coalition to be seen by the Iraqi people as liberators, rather than invaders. Efforts to seize control of the port, however, came with operational risks that challenged pilots and aircrews alike.

During the late evening hours of 20 March, MAG-39's commanding officer, Colonel Richard W. Spencer, and the 42 Commando Royal Marines' commanding officer, Lieutenant Colonel F. H. R. "Buster" Howes, finalized preparations for an aerial assault mission they had rehearsed since February. The plan called for a helicopter insertion of a unit of 550 British troops and tactical vehicles onto the al-Faw Peninsula. Based on updated intelligence data, enemy resistance was stronger than anticipated on the peninsula, and grave concerns for sabotage warranted an earlier execution of the mission. Navy SEAL teams led the assault and seized critical oil nodes earlier in the evening, paving the way for the British commandos. Before the Royal Marines were inserted onto the peninsula, a section of Cobras cleared the zone for the troop helicopters.

Leading from his command-and-control Huey, Lieutenant Colonel James R. Braden, commanding officer of HMLA-169, was the mission commander on 21 March. The 44-aircraft mission centered on the insertion of the Royal Marines from 42 Commando by a composite helicopter unit comprised of CH-46Es, a CH-53E, AH-1Ws, and UH-N1s.[21] The landing zone was about a 10-minute helicopter flight from Kuwait. Before the troop insertion, an Air Force Lockheed AC-130 Spectre gunship spent extra time on target firing into the landing zone.[22]

Just after takeoff from Kuwait, Colonel Braden's aircraft experienced a severe loss of power in one engine, forcing an immediate landing, while the rest of the aircraft continued on with the mission. Braden's aircrew, along with Colonels Spencer and Howes, quickly moved to the standby Huey and raced to rejoin the mission. Once airborne, their Huey flew into a thick blanket of black clouds that

Photo courtesy of 15th Marine Expeditionary Unit

Maj James B. Hanlon and crewmembers from HMM-161 flew primarily escort, transport, and assault support missions. During Opening Gambit, as an integral part of the 15th MEU, squadron pilots encountered some of the worst flying conditions they had ever experienced.

lingered between 1,000 and 2,000 feet. The winds were light at three to four knots from the northwest, but visibility was severely distorted because of smoke and dust from the burning oil fields; even night-vision goggles were useless. A sketchy radio transmission reported that the first wave of CH-46Es suffered a downed aircraft on the northern Kuwaiti border near Highway 801, just south of Umm Qasr. Initially believing the helicopter had experienced maintenance problems, the mission commander forged ahead into the darkness. Visibility rapidly deteriorated to the degree that flight crews could no longer see objects directly in front of them, further complicating an already congested airspace filled with 44 aircraft flying in circles over the al-Faw Peninsula. Colonel Braden vividly recalled the conditions during the evening launch: "It was invisible on the goggles, but as you entered the dust, you couldn't see because it was black. It put you into nearly instrument conditions. We had strict 3d MAW weather criteria to abort a mission given set conditions, if visibility was less than 1,000 feet, so when our flight conditions reached the limits, I called for a weather abort."[23]

During the return flight, Colonel Spencer received reports confirming that the downed CH-46E had actually crashed near the Iraq-Kuwait border, resulting in casualties for HMM-268; all 12 on board were killed, including 4 of the U.S. Marine aircrew (2 pilots and 2 crew chiefs) and 8 Royal Marines. Spencer, a Desert Storm veteran, made the difficult decision to abort the entire mission after the first wave of aircraft endured dete-

Military personnel assigned to patrol the harbor at Umm Qasr, Iraq, provide port security using an 11-meter rigid hull inflatable boat (foreground) and a 25-foot motorboat while U.S. Marines and British forces secure the nearby area.

Photo by Sgt Jeremiah Johnson, USA. Defense Imagery 030326-A-3978J-012

Photo by SPC James P. Johnson, USA. Defense Imagery 030327-A-4543J-007

Two Kuwaiti oil well control specialists prepare to extinguish a raging oil well fire at the Rumaylah oil field in Basrah Province.

riorating weather conditions. All aircraft were ordered to return to base. Having personally flown through the layers of black dust wearing night-vision goggles and not being able to see objects in front of him, he fully realized that most of his pilots were likely suffering from severe vertigo.[24] Conditions of this nature presented critical safety concerns and placed aircrews, passengers, and aircraft at grave risk.

Immediately after the colonel landed in Kuwait, he offered condolences to those who had congregated inside the Royal Marine command post. The somber period would not fade until after he personally visited the crash site a few days later. Despite the losses, the air campaign continued, and the aircrews began supporting outlying operations again, primarily with RCT-7, fully engaged in the march to Baghdad.[25] By late afternoon, the weather improved, and a smaller composite unit of British helicopters successfully landed the troops onto the peninsula.[26]

Meanwhile, HMM-161, from the 15th MEU, launched a composite transport and escort mission consisting of eight CH-46Es, five CH-53Es, and three UH-1Ns with augmentation by four CH-53s from HMM-465. The aircrews rallied at Landing Zone Eagle, while the remainder of the squadron assumed an additional responsibility at Camp Commando, Kuwait, as the I MEF tactical aircraft and personnel recovery team. At 2335 on 20 March, a company of Marines departed on board the first aircraft, commanded by Major Robert V. Boucher. Lieutenant Colonel Steven D. Peters was the assault flight leader, and Major James B. Hanlon flew as the escort flight leader.

After the first helicopter landed, the Marines began to take fire, so the second wave returned with two UH-1N escorts that provided additional close air support as troops were inserted. The CH-53E heavy transport helicopters used their external lift capability to transport four Humvees that were dropped on the eastern side of the Khawr al-Zubayr River across from the port facility. The mission was successfully completed at Umm Qasr with assistance from Royal Navy mine hunters that cleared the waterways of explosives.[27]

Major Hanlon had been flying Hueys for more than a decade during squadron deployments and as an instructor with MAWTS-1. On the evening of 20 March, less than 24 hours after the air campaign began, weather conditions complicated the night mission. Despite Hanlon's flight hours and instructor experience, he recalled,

> Flying conditions were horrible and especially at night. . . . Most of Iraq was featureless terrain, so when the moon didn't light the evening sky, it was very hard to discern between the ground and sky horizons. Further complicating conditions, the blowing dust and sand reduced sight through the night-vision goggles. Given the conditions, it was extremely difficult to keep oriented and flying. At times, our pilots had no option other than to fly the mission and either return to base or land in a safe spot until the sun rose.[28]

Airborne Direct Air Support Center Overhead

On 21 March, the airborne direct air support center—three Marine KC-130s outfitted to support aerial command-and-control missions, mirroring the mission performed by the Air Force's E-3C Sentry—began continuous coverage that lasted throughout the remainder of offensive operations. Although not as sophisticated as the Sentry, the KC-130s provided 24-hour aerial battlefield coordination operations and proved to be an invaluable asset. This was particularly true in maintaining communications with low-flying aircraft, especially helicopters that depended on line-of-sight radio transmissions.[29] In past operations, helicopters were often difficult to monitor because aircrews typically flew multiple close air support missions for several hours, landing briefly only to refuel and rearm. The airborne platform enabled direct communications with helicopters that were relayed to the TACC.

Although not routinely used in past operations because of system interface complexities and lack of available KC-130s, Colonel McFarland envisioned the need for the aerial command-and-control capability for this campaign and lobbied for the configuration. His confidence in the concept was based on improved communications technology, greater system reliability, and dedicated KC-130s. The adapted Hercules, based at Shaikh Isa Air Base in Bahrain, often operated from FOB Joe Foss in Kuwait, and each direct air support center (airborne) configuration was declared fully mission capable by late February.[30]

As the Iraqi skies filled with Coalition aircraft, the direct air support center—both main and airborne—managed the airspace that allowed 3d MAW aviators to support Marines on the ground. Just as Marine Corps sorties were visible at the joint force air command center level, the I MEF commander, General Conway, recalled days during combat hostilities when multiple U.S. Air Force bombers—Boeing B-52 Stratofortresses, Rockwell B-1 Lancers, or Northrop Grumman B-2 Spirits—were carrying up to 60 joint direct attack munitions, flying to support Marine air-ground task force operations as well as Air Force A-10s and British aircraft.[31] Air Force and Navy pilots used the Marine Corps command-and-control system and conducted joint missions with Marine tactical air controllers guiding them to their targets.

Aircraft Readiness

On 21 March, 3d MAW reported a fixed-wing full mission capable readiness rate of 69.9 percent and a corresponding partial mission capable readiness rate of 86.5 percent with secondary or backup systems. The variance was attributed to maintenance and parts discrepancies that were pending repair, procurement, or replacement. The fixed-wing readiness rates encompassed different aircraft types and models and included Harriers, Prowlers, three models of Hercules (F, R, and T), and two models of Hornets (C and D). For helicopters, the full mis-

Maj Alfred J. Croft Jr. was the officer-in-charge of the direct air support center (airborne) air controllers who flew 24-hour command-and-control operations in specially equipped KC-130 Hercules during the 1st MarDiv's march to Baghdad.

Photo courtesy of Marine Air Control Group 38

sion capable readiness rate was 76.1 percent, while the partial mission capable readiness rate was 90 percent. The UAV squadron reported rates of 78 percent and 88.4 percent, respectively.[32] The rate calculation measured squadron aircraft operability, which typically fluctuated based on several variables and trends, such as aircraft flight hours, component failures rates, parts availability, maintenance turnaround times, and aircraft modifications.* Given the age of the Corps' helicopters and KC-130s, the wing's rates were impressive, which was a direct tribute to the repair, procurement, and management efforts of the aviation and logistics squadrons.

To foster communications among the air wing, group, and operational flying squadrons, 3d MAW formed an aviation logistics operation center as part of the TACC. The new facility became a hub of activity as logisticians expedited critical demands. One of the biggest concerns was stockpiling enough aviation jet fuel (JP8 and 110LL) at air bases in Kuwait and at forward sites in Iraq to meet future demands. The center hosted a website that linked multiple internal and external agencies to provide an online forum for tracking real-time replacement parts and aircraft status. As the center's ability to juggle logistical requirements grew, it became a critical liaison point with the Marine air-ground task force's 1st Force Service Support Group and the Marine Logistics Command at CentCom to ensure delivery of supplies ranging from clothing to ammunition and repair parts.[33]

Relief-in-Place

On 23 March, the 1 (UK) Armoured Division achieved a major milestone when it assumed control of the battlespace near Basrah.[34] During an expeditious 12-hour relief-in-place south of the Saddam Canal, 1st MarDiv's RCT-5 and RCT-7 transferred their area of responsibility to the British 16th Air Assault and the 7th Armoured Brigades.[35] The transition featured extensive enemy contact. Numerous Iraqi vehicles were destroyed and approximately 300 enemy combatants were captured. The most noted Iraqi captive was Saddam Hussein's presidential advisor for scientific and technical affairs, Lieutenant General Amir Hamudi al-Saadi. Subsequent battle damage assessments and personal interviews by the 1st MarDiv with enemy prisoners of war captured during the attack revealed that a brigade-size force had been waiting for the Marines. The presence of this force further justified the earlier-than-planned night border crossing three days earlier.[36]

Although U.S. and Royal Marines encountered pockets of resistance, the Iraqis sabotaged only 9 of approximately 500 oil wells in the Rumaylah complex.[37] British forces were then placed to protect the eastern sector of Iraq south of Baghdad while the U.S. Army and Marines continued their journey north. Before I MEF could declare Opening Gambit a success, the Marines had to cross the Euphrates near an-Nasiriyah.

"Conga Line" at the Euphrates River

This next phase of the journey centered on 1st MarDiv splitting its forces along parallel avenues of approach toward Baghdad with RCT-5 and RCT-7 moving north on Highway 1 and with RCT-1 traveling on Highway 7. This required Task Force Tarawa to move through 1st MarDiv's area of operation, allowing a safe passage of lines over the Euphrates River, while General Brim's 1 (UK) Armoured Division would remain near Basrah to retain control of the southern oil fields and oversee humanitarian relief efforts.[38] Coalition forces had encountered minimal enemy resistance; however, this would change drastically as I MEF approached the river crossings.

For months, 1st MarDiv planners had discussed various options with field engineers about opportune crossing sites over the Euphrates. The challenge was to determine which one of three main roads would provide the most viable passage over the river while also avoiding travel through the center of an-Nasiriyah. Although Highway 1 was the westernmost road that linked Basrah in the south to Baghdad in the north, it initially seemed an unlikely option because it was still under construction and clearly the least developed road. Highway 8 roughly parallels Highway 1 south of the river, and both angled to the north-northwest toward Baghdad. Just south of an-Nasiriyah, Highway 7 began as a branch off Highway 8 heading almost due north, crossing the Euphrates, and passing through the city to al-Kut. After further analyses, Highway 7 appeared to be the optimal

---

*A well-established measure within the aircraft logistics and maintenance community, each aircraft type had a mission-essential subsystem matrix that identified which aircraft systems were required to conduct a particular mission. If a system could perform all missions, it was labeled as full mission capable; if it could only perform some of them, it was considered partial mission capable.

Photo by Cpl Mace M. Gratz. Defense Imagery 030322-M-IA555-012

A convoy of light armored vehicles from the 1st LAR travels along Highway 1 on the march to the Euphrates River. Numerous unit convoys stretched for miles and created vulnerable backlogs of vehicles and troops waiting to cross the river.

choice for a rapid route to Baghdad. Choosing it, however, would violate a basic principle of war—doing what the enemy expects.

The road to Baghdad options presented a dilemma between choosing an easier yet more predictable path that Iraqi military units and street fighters would most likely defend or selecting a more difficult physical route that coincided with basic war principles.[39] The Marine Corps Intelligence Agency at Quantico, Virginia, had recommended the Highway 1 option for the main effort, with RCT-5 in the lead followed by RCT-7 because analysts believed defensive positions along this route would be minimal. The agency's rationale assumed the enemy would not expect Coalition forces to use an unfinished road marked with uneven surfaces and deep ruts as a major avenue of advance. Therefore, access to Baghdad along Highways 1 and 7 would provide the Marines with two key advantages—minimized vulnerability for a consolidated enemy attack and eliminating a possible bottleneck at the Euphrates River that would jeopardize speed and surprise.

As a result, RCT-5 and RCT-7 would advance to Baghdad on Highway 1 for the next phase of the journey, avoiding major contact with the enemy, bypassing the *Baghdad Republican Guard Division* in al-Kut, and further isolating the Iraqi *III* and *IV Regular Army Corps*. Meanwhile, RCT-1 would follow Highway 7 alongside Task Force Tarawa. By 22 March, forward elements of I MEF had traveled about 100 miles north of the Kuwait-Iraq border and were quickly approaching the outer limits of an-Nasiriyah to begin the bridge crossing. Given minimal enemy engagement after moving into Iraq, a steady flow of trucks, assault amphibious vehicles, tanks, tactical equipment, and the Marines of RCT-5 and RCT-7 converged west of the city and formed multiple columns that waited to cross the Highway 1 bridge. Major traffic jams slowed tactical movement, but also caused uneasiness among commanders and senior staff noncommissioned officers. The bottleneck afforded the enemy an opportunity to deploy chemical weapons or other weapons of mass destruction against the massive gathering of ground forces in a relatively small area. General Amos described the scene best: "It was like a conga line that stretched for miles and miles."[40]

Chapter 9

Battling for the Bridges of an-Nasiriyah

A City Ruled by Fear

Although an-Nasiriyah was well marked on Coalition maps as a crossroad of three major highways on the way to Baghdad, it was an unfamiliar name from a distant land that had little significance for most Americans—until 23 March 2003. That day, journalists began reporting about an urban firefight that had ignited between Marines and Iraqi forces. It then became known as one of the bloodiest fights by Marines during the march to Baghdad that was marked by repeated demonstrations of close air support and medical evacuation missions—a genuine testimony of Marines taking care of Marines in the heat of battle.

Once a key military and economic player as Iraq's twelfth largest city, an-Nasiriyah had deteriorated into a neglected urban area. Saddam's regime purposely allowed the city's economic, cultural, and physical infrastructure to degrade as punishment because the Shiite-dominated population of more than 500,000 opposed his government after the Gulf War. The city was held not only by the *11th Infantry Division*, but also by party loyalists, the *Saddam Fedayeen*, and the Baath Party militia. These paramilitary groups were poorly equipped and trained, but they served an internal policing function for the regime via intimidation, including murder, to discourage the population from supporting anti-Saddam or Coalition forces. In the event, they fought with an intensity and brutality greater than the military forces defeated in southern Iraq.[1]

Bisected by the Euphrates River and bordered on the north by the Saddam Canal, the landscape around the city was extremely marshy, which precluded a mechanized terrain crossing or even a river fording. For RCT-1 and Task Force Tarawa, navigating to Baghdad from an-Nasiriyah up Highway 7 required two water crossings: one over the Euphrates in the south and the other over the Saddam Canal in the north. There were two bridges across each, marking the corners of a rectangle.

In the midst of a paralyzing sandstorm, RCT-1 Marines corral enemy prisoners of war and suspected fedayeen captured after a firefight north of an-Nasiriyah, Iraq.

Photo by Cpl Mace M. Gratz. Defense Imagery 030325-M-4779G-052

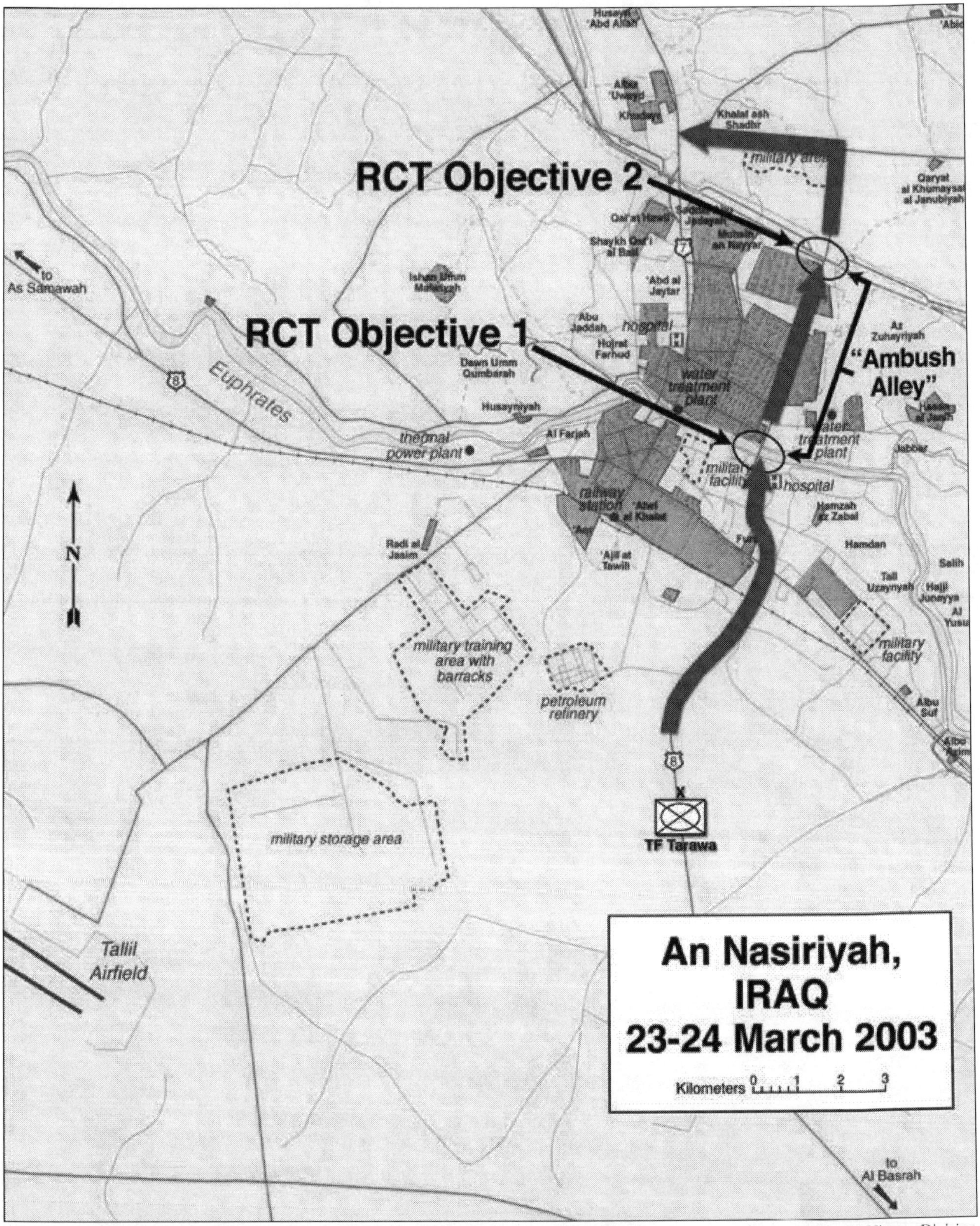

Adapted from a Central Intelligence Agency map by Marine Corps History Division

An-Nasiriyah rests on the edge of the Arabian Desert and is at the crossroad on the route to Baghdad with the intersection of three major highways—1, 7, and 8. The Euphrates River, flowing west to east, bisects the city center. The Saddam Canal runs parallel to the river on the northern edge of the city.

Photo by LCpl Bryan J. Nealy. Defense Imagery 030401-M-5455N-003

An aerial view of a main bridge in northeastern an-Nasiriyah, where intense fighting occurred between U.S. Marines of Task Force Tarawa and Iraqi soldiers and paramilitary. Securing the bridge became vital to ensuring logistical movement as a main supply route to forces positioned in northern Iraq.

The river bridges were called the southwestern and southeastern bridges, and the canal crossings were deemed the northwestern and northeastern bridges. The three-mile stretch of road that connected the eastern crossings and skirted the edge of the city was known by U.S. Army planners as "Ambush Alley" because a convoy traveling this route would be vulnerable to enemy sniper and small-arms attack.[2] Regardless, Marines marched north toward the city with every intention of bypassing it on the way to Baghdad. On the evening of 22 March, elements of Task Force Tarawa camped near the cloverleaf intersection of Highways 7 and 8 just south of an-Nasiriyah. Gunnery Sergeant Kevin Berry, the operations chief of 1st Battalion, 2d Marines, recalled the pause before battle: "It was like most other previous nights. We got a bit of sleep. We had been eating military prepackaged meals-ready-to-eat for over 40 days plus Copenhagen and the first sergeant's coffee for dessert."

BGen Richard F. Natonski, commander of Task Force Tarawa, speaks with an embedded journalist after Marines secured the bridges of an-Nasiriyah. Following an intensive firefight between U.S. forces and civilian clad enemy combatants, the battle guaranteed multiple crossing sites over the Euphrates River, expediting the march to Baghdad.

Photo by LCpl Bryan J. Nealy. Defense Imagery 030329-M-5455N-001

Seize All of the Bridges

While control of both western and eastern bridges made sense based on the rationale that multiple crossing points would thwart a potential bottleneck at the Euphrates, the western route would force convoys through the densest, most built-up areas of Ambush Alley. Adding to concerns, General Conway had received reports that the surface condition of Highway 1 was in terrible repair with deep ruts and gullies, thus jeopardizing the ability to use it as the sole advance and supply route for RCT-5 and RCT-7.[3] Despite General Conway's guidance that Marines were to bypass urban areas, Task Force Tarawa would have to make their way up Ambush Alley.[4] Anticipating that Task Force Tarawa would most likely execute a "be prepared to" mission, General Natonski was not surprised when he received a verbal order on 22 March followed by Fragmentary Order 0 17-03 to capture the eastern bridges.

As an avid reader of military history, General Natonski expected that his command might encounter pockets of resistance, possibly leading to urban combat. Further supporting the potential enemy threat were reports that Saddam's cousin and one of his most trusted loyalists, Ali Hassan al-Majid al-Tikriti, was posted near an-Nasiriyah and ready to defend southern Iraq with chemical weapons against approaching ground units.* Concerned that an-Nasiriyah could become a distraction and slow the momentum of 1st MarDiv, General Natonski's mission gained a greater sense of urgency, especially since RCT-1 was trailing close behind on Highway 7 and was eager to move on its route to Baghdad.[5]

Task Force Tarawa units moved into attack positions about eight miles south of the southeastern bridge, maintaining satellite and secure voice communications with the command post at Jalibah. As they began executing their attack plan, chatter flooded tactical radio nets with reports of an enemy ambush of a U.S. Army convoy that had strayed through the city to the north of Task Force Tarawa's position. This caused concern for two reasons: fire support coordination plans were made with no knowledge of any Coalition units in the town, and enemy resistance was still expected to be light. Subsequent reports confirmed that an 18-vehicle convoy from the U.S. Army's 507th Maintenance Company had taken a series of wrong turns into an-Nasiriyah and Ambush Alley. Commanded by Captain Troy K. King, USA, the unit mostly consisted of cooks, mechanics, and computer technicians who lacked the basic infantry skills necessary to defend themselves against ambush or attack. The commotion of a tactical convoy driving through the center of the city caused an immediate reaction from a mix of Iraqis from the *11th Infantry Division*, fedayeen loyalists, and local paramilitary street fighters. Initially, enemy forces began attacking the convoy with small arms. The organized enemy forces wore uniforms, whereas most of the random street fighters wore civilian clothes and ducked in and out of public facilities and residential homes in an attempt to blend among civilians, making it difficult to return fire. As Iraqi fire intensified, some of the U.S. soldiers dismounted their vehicles, while others remained near their vehicles trying to repel attackers, which attracted even more fire, since the enemy perceived a Coalition weakness.

A firefight quickly developed and lasted about 60–90 minutes, drawing Colonel Ronald L. Bailey's RCT-2 into the skirmish. At that time, his 1st Battalion, 2d Marines, became fully engaged, attacking through a barrage of enemy artillery, mortar, and machine-gun fire to aid the Army unit while obtaining medical assistance for wounded soldiers. Embedded journalists reported that an enemy ambush had resulted in the deaths and capture of several soldiers, including Private First Class Jessica D. Lynch, USA. Major William P. Peeples, a reservist and a city planner from Avon, Indiana, was one of the first to meet U.S. Army soldiers on a dirt road south of the Euphrates River crossing. As commander of Company A, 8th Tank Battalion, Major Peeples recognized the severity and proximity of the fight and immediately coordinated with headquarters for rescue and casualty evacuation assistance. Still fighting to gain control of the eastern bridges and prevent their possible demolition, Task Force Tarawa dealt with the disruption, drastically changing the original attack plan and threatening a planned seizure of all the bridges at 1000.

In an interview, General Natonski later recalled the events on 23 March:

> I understood RCT-2 had encountered some small-arms fire when they moved out of their attack position south of an-Nasiriyah, but when I learned they were clearing buildings on both sides of the road . . . I was concerned at the slow pace of their movement, fearful they would never get through the city in daylight. The flexibility that command-and-control [UH-1N] aircraft permitted me to

*Nicknamed "Chemical Ali" because he used poison gas against the Kurds in northern Iraq during the Gulf War, al-Majid was a concern to U.S. commanders because they thought he would use similar weapons against them without hesitation.

move around the battlefield greatly improved my situational awareness and facilitated operations that day as I flew a couple of times from the regimental position in Jalibah to an-Nasiriyah to personally see the progress. Spotting four wounded soldiers in the midst of a Marine CH-46 medical evacuation, I subsequently discovered they were from the ambushed Army unit. My thought was twofold. We had to rescue the soldiers, yet at the same time secure the bridges while we had the combat force available and daylight. We later learned from an Iraqi executive officer of the *23d Brigade* that they had been emboldened that the U.S. Army convoy had been attacked so easily . . . so the Iraqis thought they could claim an easy victory.[6]

General Natonski flew back to Jalibah where he briefed General Conway by secure telephone. Thirty minutes later, Natonski returned to an-Nasiriyah and gained more insight into RCT-2's attack with artillery support and air wing Harriers and Cobras, providing aggressive close air support. Meanwhile, 1st Battalion, 2d Marines, had encountered terrain, communications, and logistical hurdles that impacted its assigned roles in securing the bridges. The attached Company A, 8th Tank Battalion, was still refueling and rearming after the recovery of the Army maintenance convoy. As events unfolded, Company A held the southeastern bridge, allowing follow-on forces to cross the Euphrates, including Company B along with battalion headquarters, which had become bogged down in mud along its assigned route.

Meanwhile, Company C took the initiative and assumed the lead, rushing to secure the northeastern bridge over the Saddam Canal; however, the unit was unable to inform or coordinate its revised role to the battalion headquarters or regimental command staff because of a breakdown in communications. As the company pressed north, Iraqi regular and irregular forces appeared from all directions, popping in and out of buildings and firing from rooftops. Because the enemy wore civilian clothes and hid among civilians, Marines could not return fire because the rules of engagement prohibited combat unless the enemy was positively identified, and the rules also sought to minimize damage to the public infrastructure.[7] The situation worsened when a rocket-propelled grenade (RPG) hit an armored assault vehicle, engulfing it in flames. The explosion killed and seriously wounded several Marines from Company C, who repelled enemy forces as best they could while requesting medical evacuation assistance, artillery, and close air support.

Close Air Support and Forward Air Control

In accordance with Marine Corps doctrine, the tactical air control party and air officer were with 1st Battalion headquarters, and two FAC teams were assigned to the assault companies (Company A tanks and Company B), leaving one rifle company without an FAC (Company C). An attack pilot had to talk with an FAC or the direct air support center before engaging or altering altitude.[8] Such coordination measures ensured centralized command yet disciplined procedural control of aircraft—a hallmark of the Marine air command-and-control system.

During the early morning hours of 23 March, Major Scott S. Hawkins, the FAC assigned to Company A, 8th Tank Battalion, briefed the operational situation to this unit as it began moving northward toward an-Nasiriyah to seize the bridges:

> Here is our plan. We anticipate little or no resistance from *Republican Guard* and fedayeen forces until we advance north of the city, where we could expect artillery, tanks, and armor. We're going to be a channelized mechanized force approaching a fortified city. Tanks are the lead element of the lead battalion for [Task Force] Tarawa. Okay, I want Cobra over-launch.

As Major Hawkins's unit approached the city, he noticed a lone vehicle approaching from the opposite direction that quickly turned around in the middle of the road. Suspecting that it was a military truck, he observed a man with a white pickup literally picking up children and shoving a woman into the back of this truck as fast as he could. As Major Hawkins recounted,

> This isn't good. Twenty seconds later small-arms fire started, followed by mortars. We had driven in between two farmhouses. Why they decided to take on the lead tanks was a mystery. It was pretty dumb. They could have let several tanks pass, then they would have had plenty of soft-skin targets, but they

> decided to take us on first. The initial reaction was what tankers do, they turn their nose into the fight and spread forward. The tanks just went off the road and pushed right up to the houses.[9]

Entering the fight, the tanks responded with .50-caliber rounds, transitioning to a troops-in-contact formation near a garbage dump that had attracted hundreds of huge black flies. Nearby, a mangled mess of power lines hung about 40 feet off the ground—a helicopter's worst nightmare when trying to land. Meanwhile, a division of three MAG-29 AH-1W Cobras from Lieutenant Colonel Jeffrey M. Hewlett's HMLA-269 arrived. During this time, the ultra high-frequency radio belonging to the 1st Battalion air officer was inoperable, and backup communications between forward units and headquarters staff was intermittent. Regardless, Major Hawkins communicated directly with pilots, other FACs, and the direct air support center to gain situational awareness. Call signs listed on the daily air tasking order became cumbersome and confusing, so pilots and controllers used aircraft types to retain control of a saturated airspace and depended on the familiar voices of FACs and direct air support center air controllers to ease confusion as pilots rotated and exchanged mission information. In retrospect, Hawkins summarized the day's events as follows: "The Cobras were working, and we just wanted to keep them overhead. Our biggest fear, shared by Marines on the ground, was the Cobras were going to be diverted elsewhere, but they [were] not."[10]

Three Types of Close Air Support

Dating back to the 1920s in Nicaragua, close air support techniques gained recognition during the Banana Wars in Latin American. In these conflicts, which had one of the first reported instances of the technique, the benefits of using aircraft for offensive operations were demonstrated when Marines on the ground directed pilots toward enemy positions by using signal panels.[11] From that point, technological advancements in aircraft capabilities in conjunction with aircrew and controller training, radar, and improved communications continuously refined the techniques. Reliance on the support grew during subsequent conflicts through World War II, Korea, and Vietnam, demanding that protocols become more clearly defined to the point that it became a published doctrine.

In the early 1970s, former Marine Corps Commandant General Robert E. Cushman summarized the sentiment toward integrating the air-ground team: "I still need Marines who can shoot and salute. But I need Marines who can fix jet engines and man sophisticated radar sets, as well."[12] Because the Marine Corps prides itself on air control procedures, these types of close air support appeared easier for its pilots and controllers to adapt to than those of the Air Force, even if the direct air support center had joint service aircraft in its airspace with pilots wanting to fly Marine Corps missions.[13]

With the new joint close air support publication still in draft status, the Marine Corps defined three types of close air support:

- Type I is the most restrictive, requiring the FAC to see the attacking aircraft and the target. This reduces the potential for friendly fire incidents or an unintended attack on noncombatant targets.
- Type II requires that the FAC see either the target or the attacking aircraft for individual strikes.
- Type III is the least restrictive and does not require the FAC to see either the target or the attacking aircraft for multiple strikes. Target identification and accurate target strike are the responsibility of the pilot, which imposes a higher risk on friendly ground forces.*

*Joint doctrine as found in Joint Publication 3-09.3, *Joint Tactics, Techniques, and Procedures for Close Air Support (CAS)*, classifies close air support by three types: type I, type II, and type III.

Friendly Fire Incident

During the fight for the bridges, Task Force Tarawa experienced a friendly fire incident during a close air support mission flown by U.S. Air Force A-10s. Company B's FAC, Captain Dennis A. Santare, conveyed a sense of urgency using the "guard" or emergency frequency, which indicated "troops-in-contact." Similar to an emergency 911 call, temporary pilot discretion was warranted when under fire as long as the pilot was under positive control of an air controller from a controlling agency, which allowed battlefield interpretation of the rules of engagement. A request for help by Company B's FAC resulted in the immediate response of two Warthog pilots checking in with the direct air support center and offering assistance.[14] Concerned about casualties, the FAC, with concurrence of Company B's commander, authorized the two A-10s to use type III close air support without being aware that Company C had become the forward unit at the northeastern Saddam Canal Bridge. This particular mission required the aviators to acquire the target and drop ordnance independently of a controller (doctrinally, this scenario required authorization by a battalion commander). The A-10 pilots spent about 15 minutes coordinating and fine-tuning their on-call mission with the controller before dropping their ordnance.[15]

As requested, the A-10s conducted multiple air strikes, dropping eight MK82 500-pound general-purpose bombs and launching three AGM-65 Maverick air-to-ground missiles as well as strafing with their 30mm cannons. They attacked tracked vehicles that were heading south on Highway 7 toward Task Force Tarawa, believing they were the enemy as no friendly forces had been reported that far north.[16] In fact, the pilots were attacking elements of Company C. By early evening, the company reported that it had lost 18 Marines to both enemy RPG hits and the ordnance from the A-10s.[17]

Although seizing the bridges marked an end to battle, it also marked the beginning of an exhaustive investigation into this friendly fire incident.* As expected, CentCom appointed a panel led by an Air Force brigadier general with a Marine colonel advisor. The report, including a comprehensive collection of documents and witness statements, was completed in 2003 with a version available for public review a year later. Attributing the incident to several factors, the findings highlighted a breakdown in radio communications; a lack of updated enemy intelligence; battlefield confusion or the "fog of war"; and a change in the attack, which placed Company C as the lead unit without situational awareness by higher headquarters.

The main concern, however, was that Company B's FAC authorized type III close air support without first obtaining permission from the battalion commander. The report noted that the ground FAC had not acted recklessly or negligently but was motivated by a desire to assist Marines in combat. The report also concluded that of the 18 Marines lost that day, 8 were attributed solely to enemy fire.[18] Although the final report cited no action against the A-10 aircrews or any battalion personnel, it did recommend taking administrative or disciplinary action against the FAC for authorizing type III close air support without the battalion commander's approval. Interestingly, the flight tapes from the cockpit voice and video recorders were never recovered from either of the Warthog pilots who flew the mission. After internal review, the Marine Corps sought no action against any of the Marines involved in the incident.[19]

View From Above

On the morning of 23 March, news circulated quickly over crowded radio frequencies and by word of mouth at Jalibah airfield that Marines were fiercely engaged with enemy forces near an-Nasiriyah. The aircrews scanned radio frequencies, listening to bursts of air traffic dialogue among airborne and ground FACs, trying to piece together situational awareness. Regardless, during the early morning hours, F/A-18D Hornets, AV-8B Harriers, UH-1N Huey gunships, and AH-1W Cobras flew toward an-Nasiriyah and checked in, hoping to answer requests from FACs and the direct air support center for immediate close air support missions. Ground forces were in trouble, and Task Force Tarawa needed attack aircraft overhead.

Providing 24-hour coverage over an-Nasiriyah, aircrews devised a rotation in which aviators contacted air controllers, gained tactical situational

*DoD Instruction 6055.7, *Accident Investigation, Reporting, and Record Keeping*, requires investigation of friendly fire incidents, as completed by the U.S. Air Force for the incident at an-Nasiriyah. The incident and subsequent investigation generated internal review among joint service commands regarding how to prevent similar occurences in the future. As a result, Joint Publication 3-09.3 for close air support was revised, and funding requests to procure additional automated tracking systems to distinguish between friendly and enemy forces were approved.

Photo courtesy of Col Michael D. Visconage

Fixed-wing attack pilots flew countless high-altitude missions, searching and striking enemy targets in advance of ground forces. In this photo, an F/A-18 Hornet pilot from VMFA(AW)-121 conducts such a mission, further reinforcing the aviation combat element role as a separate maneuver element.

awareness, hit targets (including tanks, mortar positions, recoilless rifles, and small-armed enemy positions), refueled, and returned to attack more targets. All-weather fighter-attack F/A-18Ds and AV-8Bs served dual roles, alternating as airborne FACs and aligning with UH-1N and AH-1W gunships to provide continual close air support for Task Force Tarawa. Air controllers stacked fixed-wing aircraft at different altitudes until they could assign on-call missions, while the airborne direct air support center controllers, who had just established operations a day earlier, provided oversight of helicopter operations.

Hornets typically flew strike coordination and armed reconnaissance missions, since the aircraft was a fighter/attack aircraft, but the Litening II targeting pod made it appealing to use Harrier attack jets in these roles as well. Captain Mikel R. "Mike" Huber, a MAWTS-1 instructor augmenting 3d MAW, experienced firsthand the success of the Litening II and aircraft capabilities that had been developed before the war. Drawing from his perspective at an-Nasiriyah, he recalled:

> The targeting pod provided a pretty decent sanctuary. You felt good from 15,000 to 20,000 feet when you can look down with the sensor and destroy targets, day or night, while the bad guys [are] having a tough time seeing or engaging me with antiair missiles. When the airplane was configured with this sensor, a pilot's situational awareness in-

LCpl Matthew T. Riddle from MWSS-373 guides an AH-1W Cobra after it refueled at Jalibah Air Base and is about to return to an-Nasiriyah. Aviators provided continuous air support coverage, rotating from Jalibah, a major forward operating base, until the Marines secured the city almost a week later.

Photo by LCpl Christopher H. Fitzgerald. Defense Imagery 030323-M-2237F-006

It Sounded like Getting Hit with a Baseball Bat

Captains Craig H. Streeter and Matthew R. Shenberger had just finished scrolling through the air tasking order at Jalibah airfield around 0630 local time, 23 March, when they overheard vague reports of troops-in-contact and possibly an engagement involving Marines near an-Nasiriyah, about 30 miles north of their position. They finished their preflight checks and understood a simple one-word command: "Launch!" Streeter operated initially as an airborne FAC and worked with Major Scott Hawkins to help manage the airspace by stacking F/A-18s at different altitudes to await assignment with Cobra gunships. Streeter coordinated a flight of two Marine Hornets using a forward-facing infrared sensor and precision munitions that could destroy targets inside a tree line with pinpoint accuracy. Major Hawkins provided immediate feedback to the pilots: "The Marines are loving it."

When Captain Shenberger joined the fight, he had already flown more than 10 hours and recalled the situation about midway through his long day:

> Wow, this is pretty serious because we hadn't even breached the city yet and this was happening. My day got interesting about 30 minutes after the Company B tanks got bogged down and Company C crossed to secure [the] bridge. We were still trying to assess where the forward line was located, so from my perspective I was just trying to get eyes on target. Suddenly, it sounded as if my helicopter was hit with a baseball bat on the canopy. It was [a] loud bang and I noticed something fall onto the ground. It looked like a red tracer that was burning out, it didn't seem normal, but my wingman didn't notice anything missing from my aircraft. I was flying fine; my gauges, instruments, and engines were working fine. I didn't have the luxury to go land somewhere and inspect it a little more closely, so we continued looking for targets on both sides of Ambush Alley. We noticed people fleeing houses and buildings, so it was becoming more difficult to distinguish civilians.[20]

Capt Matthew R. Shenberger (right), a Cobra pilot with HMLA-269, flew more than 10 hours during the battle of an-Nasiriyah, providing close air support for Task Force Tarawa Marines. The squadron's commanding officer, LtCol Jeffrey M. Hewlett, awarded the Distinguished Flying Cross to Capt Shenberger along with three other squadron pilots—Capts Brian Kennedy, Andrew Dyer, and Eric Buer—for their efforts during Operation Iraqi Freedom.

Photo courtesy of Marine Light Attack Helicopter Squadron 269

During an engagement, Capt Shenberger's AH-1W was hit by enemy fire. After assessing the damage and carefully checking cockpit instruments and gauges, he continued flying for a few more hours, unaware of the severity of damage to one of the rotor blades.

Photo courtesy of Marine Light Attack Helicopter Squadron 269

creases tremendously. As a result, Harriers were assigned to fly strike coordination and armed reconnaissance missions on the air tasking order along with Hornets. . . . It was something to see![21]

MAG-29, a composite helicopter group from New River, North Carolina, played a major role during the an-Nasiriyah battle. Commanded by Colonel Milstead, MAG-29 and Task Force Tarawa had jointly trained and traveled together, so aircrews naturally wanted to help as the fighting raged. Attached to 3d MAW and consolidated with West Coast squadrons, AH-1W Cobra aircrews provided continual coverage for hours, conducting close air support missions with brief landings at Jalibah for a hot refuel before returning to the city.

A constant rotation cycle worked especially well for helicopter crews, who often flew multiple missions that totaled 8–12 hours with short replenishment stops. Typically, Cobra aircrews took 30–45 minutes to fire their missiles before they flew a short distance, usually less than 10 minutes, to quickly refuel and rearm. Flying at altitudes as low as 100 feet and at speeds of 80–120 knots, aviators destroyed a mix of well-defended Iraqi targets, including tanks, armed vehicles, mortar positions, and recoilless rifles. A typical Cobra ordnance load included 2.75-inch Hydra 70 rockets, AGM-114 Hellfire missiles, and a M197 20mm cannon.

Casualties and Medical Evacuations

By midafternoon, helicopter aircrews landing amid enemy fire began transporting Marines from the an-Nasiriyah battlefield. The direct air support center coordinated medical and casualty evacuation missions with flights of CH-46s, CH-53s, and UH-1Ns into nearby combat areas to transport fallen and seriously wounded Marines to advanced medical facilities.[22] Accounting for the location of every Marine has always been a high priority and great responsibility for commanders, especially in combat environments. Once a casualty is confirmed, a team of specially trained Marines follows a detailed administrative process and protocol to ensure unit and family death notifications are performed with utmost respect for the Marine and next of kin.

In March 2003, the command staff of Marine Corps Forces Europe in Stuttgart, Germany, formed a hospital liaison team at Landstuhl Regional Medical Center near Ramstein Air Force Base that assisted in processing wounded Marines to their final destinations. The team coordinated all transportation details from the injury site, the medical evacuation out of theater, and throughout the recovery period, which meant strict patient accountability and casualty status reporting. During the first month of OIF, the hospital liaison team received 53 Marines with combat-related injuries; in the next month, the facility took in 135 casualties.[23] When family members visited the medical complex and walked through a maze of corridors and reception desks, they had no difficulty finding the liaison team's home since the Marines were the only service prominently located inside the facility.

Command and Control in a Saturated Airspace

As demonstrated during the an-Nasiriyah battles, the I MEF airspace became saturated with aircraft. Communications were jammed as primary radio frequencies became overloaded with too many voices and the tactical radio frequencies that relayed transmissions from aircrews, air controllers, and ground commanders. Regardless, aircrews forwarded information about the battle when they landed to refuel at Jalibah and other forward sites. Despite periods of interrupted communications, other factors enabled the air-ground team and MACG-38 control agencies to retain command and control of fixed-wing aircraft and, even more importantly, the attack helicopters that were flying barely 100 feet above rooftops and power lines. When radio transmissions became garbled or multiple call signs assigned on the air tasking order became too cumbersome to use, the sound of familiar voices among the air-ground team prevailed, enabling pilots, FACs, and air-ground liaison Marines to effectively communicate. The effectiveness of Marine systems to exercise centralized command and decentralized control was attributed to years of air-ground task force training exercises and real-world operations coupled with disciplined pilots and air controllers.

The events that unfolded in an-Nasiriyah reinforced the decision of Colonel McFarland, commander of MACG-38, to extend kill box communications and surveillance coverage to support the northern advances toward Baghdad, so he arranged for an early warning-and-control detachment to move forward

Blair Field Named to Honor a Fallen Marine

RCT-2 became the lead unit during the mission to secure the eastern bridges of an-Nasiriyah. Captain Jeffrey A. Vandaveer, commanding officer of Battery B, 2d Low Altitude Air Defense Battalion from MACG-38, had traveled with Task Force Tarawa ever since it crossed the line of departure. The battalion conducted mechanized and motorized combat patrols and provided perimeter security, area surveillance, target identification, and location information for mortar, artillery, and sniper teams. During the fight on 23 March, which continued for several hours, Captain Vandaveer's unit set up air and perimeter defense alongside RCT-2 and received indirect and direct enemy fire.

Photo by LCpl Andrew Williams. Defense Imagery 030825-M-UW798-015

Members of 1st MarDiv cross the flight line after exiting a Sea Stallion assigned to HMH-462 at Blair Field. Once FOBs were established throughout Iraq, Marines were redeployed using transport aircraft as an alternative to traveling in the more vulnerable ground convoys.

One day later, Vandaveer received formal notification from a platoon leader that a Marine was missing. Lance Corporal Thomas A. Blair of Wagoner, Oklahoma, a battery gunner attached to Company C, 1st Battalion, 2d Marines, had disappeared during the an-Nasiriyah battle at the northeast bridge the previous day. The captain submitted a personnel casualty report with duty status "whereabouts unknown" to group headquarters along with a wounded in action report for another Marine who was hurt in the A-10 friendly fire incident.

With assistance from Task Force Tarawa and I MEF headquarters, the air defense battalion conducted an extensive search that revealed Corporal Blair had been injured and evacuated from the battlefield. Five days later, the report was proven incorrect when his body was identified near the battlespace in an-Nasiriyah. Captain Vandaveer then completed a killed-in-action report that initiated the movement of the remains to Camp Doha, Kuwait, then to Dover, Delaware, before being sent for burial in East Joplin, Missouri. Blair was the first Marine Corps low altitude air defense battalion gunner killed in combat. General Natonski was deeply touched by the sacrifices of the young Marine and sought to formally recognize him. On 22 April, Corporal Blair was memorialized when al-Kut airfield was renamed "Blair Field."[24]

Two CH-46 Sea Knights (left) and three CH-53E Super Stallions, all from HMM-364, are parked on the ramp at Blair Field in al-Kut, Iraq.

Photo by LCpl Andrew Williams. Defense Imagery 030903-M-UW798-030

Direct Air Support and Task Force Tarawa

Major Mark D. Tobin was the officer-in-charge of the Task Force Tarawa air support element during OIF. He had enlisted as a reservist and achieved the rank of sergeant as a bulk-fueler before he was commissioned as a second lieutenant and subsequently chose the Marine Corps as a career. With a firsthand understanding of air-ground operations, Major Tobin vividly recalled events as the air support element, which mirrored the operations of a direct air support center, neared an-Nasiriyah on 23 March:

> My Marines weren't the ones being shot at, but the Marines we were supporting were. We had just picked up and moved recently, and we were setting up Task Force Tarawa. . . . [The] combat operation center . . . was in a landfill on the outskirts of town. A town was right there, about half a click [kilometer] away. We wondered why we were setting up here because it was nasty, but we did it anyway. Our senior air director, Captain Joanna L. Garcia, was in charge when we got the first call for a medical evacuation. The reports started pouring in with hints of how bad it was, so I ran over toward her to get a first-hand assessment. My natural instinct was to kind of elbow her, get her out of the way, and I'll take over, but I bit my tongue and I stood back. I'm glad I backed off because she did an incredible job and I was extremely proud of her. I'd been in a direct air support center before when a Marine had gotten shot, but this was the first time, ever, that I could sense such emotion through a radio in Marines' voices.
>
> The 8th Tank Battalion forward air controller, Major Hawkins, was desperately requesting air and we were getting multiple calls and inquires from the combat operations center side as well. It was very frustrating because we had a very hard time getting the CH-46s into landing zones for medical evacuations. Enemy fires were fierce and the "hot" landing sites were too risky to attempt. However, Captain Eric Garcia from HMLA-269 miraculously and heroically landed and evacuated wounded Marines in a hot landing zone, while respectfully loading Marine casualties. With enemy fire all around, Captain Garcia and his aircrew saved the lives of several injured Marines that afternoon.
>
> My controllers even talked to the Air Force A-10 pilots who were involved in the friendly fire incident with Company C. We turned the [A-10] pilots over to the forward air controller and, [at] that point, we were following proper protocols. For the perspective of the battle watchstanders who were on duty that shift, it was a tough couple days. It was solemn. We felt the impact of battle as air support Marines. It created a genuine sense of reality and seriousness.[25]

to Iraq. Despite initial logistical hurdles in finding transportation, the tactical air operations center moved from Kuwait to Jalibah FOB. Within three days, the detachment linked its TPS-59 and TPS-63 radars into the data link system and provided a radar picture of the air war. Given its proximity to the hub of air operations, the detachment assisted in controlling Marine KC-130 tanker missions and directing airborne direct air support center aircraft, as needed.[26] Direct air support centers on the ground or airborne managed the I MEF airspace through altitude separation—dictating that fixed-wing aircraft descend no lower than 10,000 feet, airborne FACs no lower than 5,000 feet, and helicopters as low as 100–200 feet.[27] The accelerated pace of air operations challenged even the most seasoned air controllers considering that more than 400 Marine aircraft were in theater.[28]

The collaboration of well-trained airborne FACs with ground tactical air controllers was instrumental in directing aviators to targets and assisting MACG-38 in maintaining a safe airspace. Further, the two-man F/A-18D Hornets used by 3d MAW in the aerial controller role were central to employing aviation as a maneuver element, in which the aircrews identified additional targets and directed other tactical aircraft to them.

Leveraging this, airborne FAC capabilities supplemented the demanding workload managed by the direct air support center controllers and its subordinate air support elements. Hornets were particularly welcomed by the direct air support center during high tempo peaks when the airspace became saturated because they could "pull" or divert the F/A-18s from search-and-destroy missions and redirect pilots for immediate close air support. This tactic was much more efficient than stacking aircraft until mission assignment.

In a later interview, Colonel Alles, the MAG-11 commanding officer, described the coordination and sequence of events in which the group's Hornet aircrews attacked stationary Iraqi divisions north of an-Nasiriyah, near al-Kut. Morning missions were flown to obtain aerial reconnaissance imagery for later attack missions. As a result of the success of this cycle, Iraqi units virtually disappeared ahead of the ground forces as they moved toward Baghdad. Colonel Alles assessed the situation this way:

> I believe MAG-11 may have only destroyed 10 percent of the enemy capability, but our approach caused many [personnel] to simply abandon their units and equipment. There was essentially no organized defense as Marine ground units moved up Highway 6 toward Baghdad. There was an AAA [anti-aircraft artillery] threat, but it was mostly visible at night. Ultimately, the MAG suffered no battle damage.[29]

After an-Nasiriyah

As Task Force Tarawa moved deeper into an-Nasiriyah on 23 March, Marines encountered enemy vehicles, but no trace of the enemy; the Iraqi equipment had simply been abandoned. By approximately 1800 on 23 March, Tarawa had seized both bridges, and a day later, most of 1st MarDiv was north of the Euphrates River. Although the combat was most intense for about four hours, the increased tempo continued for more than three days as the task force secured all of an-Nasiriyah. The area remained unsecure for almost a week. Fully anticipating a major vehicle convoy backlog at the Euphrates crossing, I MEF logisticians had compiled a synchronized movement of supplies, equipment, and Marines for the first of several river crossings during the march to Baghdad. Supplemented with firepower from Task Force Tarawa and close air support from 3d MAW, 1st MarDiv split its forces at Highways 1 and 7, crossing the Euphrates River at two sites—the eastern bridges in town for Highway 7 and the Highway 1 bridge to the west of town—and sustaining a momentum that exceeded projected timelines. This flow could not have been accomplished without continuous, 24-hour overhead coverage by the wing's aircrews and tireless tactical air controllers.

By 28 March, Task Force Tarawa had steadily moved its forces east of the crossing sites and expanded its control over an-Nasiriyah, eliminating remnants of lingering enemy resistance. The task force seized an enemy strongpoint at the Tykar Hospital in eastern an-Nasiriyah and discovered 3,000 chemical protective suits along with large caches of weapons and ammunition.* At the same site, Marines captured more than 500 prisoners and gathered evidence regarding the captivity of U.S. Army troops who were attacked in Ambush Alley earlier, thus improving the likelihood of finding the missing soldiers.[30]

Seizing the bridges of an-Nasiriyah granted access to Highways 1 and 7 and empowered 1st MarDiv and 3d MAW to fix, bypass, and isolate the *Baghdad Republican Guard Division*, along with the rest of the Iraqi *III* and *IV Corps*, near al-Kut. Heading north toward ad-Diwaniyah, RCT-5 and RCT-7 crossed the western Highway 1 bridge, as RCT-1 moved north along Highway 7 toward al-Kut. Despite the bloody battle to secure the eastern bridges, intelligence recommendations to use the Highway 1 crossing as the main route proved to be the correct decision, allowing 1st MarDiv to move more than 8,000 vehicles through this choke point within 12 hours.[31]

Although organized Iraqi military forces temporarily disappeared south of the Euphrates River after Task Force Tarawa gained control of the area, Saddam unleashed his fedayeen loyalists from Baghdad to intimidate the local population, forcing it to resist Coalition forces. As a result, gaining control of an-Nasiriyah took longer than expected, causing more casualties than projected and expanding the scope of operations along Highways 1 and 7, which required reinforcements. Task Force Tarawa therefore temporarily assumed operational

*The discovery of such stores, a blatant violation of the Geneva Convention, confirmed that the enemy was using the facility for combatant purposes.

Photo by Sgt Zachary A. Bathon. Defense Imagery 030330-M-7371B-004

A night-vision photo of Marines from the 24th MEU (SOC) putting on their gear before moving to a transport aircraft destined for Iraq.

control of the 15th MEU, the 25th Marines, and Task Force Yankee.

On 29 March, the 24th MEU (SOC), commanded by Colonel Richard P. Mills, began supporting Task Force Tarawa, increasing General Natonski's troop strength to 13,379 personnel.[32] Its aviation combat element, HMM-263, was temporarily attached to MAG-29 under wing command. The squadron watched over RCT-1 as it moved near Qalat Sikar Air Base and south of al-Kut north along Highway 7.* The squadron not only conducted immediate air support missions, but also provided assault support and played a key role in the movement of troops and supplies.

Throughout the weeklong period to secure the area around an-Nasiriyah, General Robling was called several times by General Natonski to personally thank the air wing for its support. As General Robling remembered,

> The last week of March was particularly ugly, and 3d MAW continued to work with Task Force Tarawa, giving them what they needed. Each night, we would get a call from Brigadier General Natonski, thanking us for what we had done that day and then projecting what aviation support Task Force Tarawa would need next. An-Nasiriyah was our first real exposure to the urban fight. It highlighted the fact that aviation can't do everything unless you level an entire area, and Marines still have to go in and clear the enemy.[33]

Photo by LCpl Bryan J. Nealy. Defense Imagery 030331-M-5455N-006

Task Force Tarawa Marines distribute donated food supplies for humanitarian relief to Iraqi citizens in an-Nasiriyah a week after securing the bridges and establishing order in the city.

Although Task Force Tarawa bore the deepest scars from the battle of an-Nasiriyah, 3d MAW shared in their loss, flying continuous close air support and evacuation missions for the dead and wounded amid fierce enemy resistance. The extended firefight stirred emotions, and, in general, Marines soon changed their opinion of the enemy when they realized they were fighting combatants who wore civilian clothing and hid behind women and children. Expecting to encounter the Iraqi *11th Infantry Division* near an-Nasiriyah, the air-ground commanders did not anticipate such fierce resistance from Baathist and fedayeen irregular forces and a kaleidoscope of paramilitary forces that reportedly included Palestinians, Pakistanis, and Chechens. Often, this enemy wore civilian clothes while waving white surrender flags, only to open fire a few seconds later. Indeed, an-Nasiriyah shed light on a much different enemy force than expected and planted fresh doubts among intelligence community analysts about initial enemy estimates.[34]

The aftermath of an-Nasiriyah made Marines of all ranks eager to reach the enemy's center of gravity: Saddam's regime in the heart of Baghdad.[35] The enemy's crude war-fighting techniques altered air-ground team tactics during the final planning effort for the Baghdad offensive, particularly regarding helicopter and urban close air support missions.[36] In April 2003, General Robling provided this perspective on the shift:

---

*The 24th MEU (SOC) had just finished its participation in a joint exercise (Iron Magic 03) in the United Arab Emirates and had its six-month deployment extended to participate in OIF.

We were seeing the 1st Marine Division taking casualties, and Marines were dying. The air wing slowed down and reassessed how we were fighting and the ground side did the same thing. It was just a low time in the fight for us, since we had experienced many successes up to this point. It may have seemed as though we went over the edge; however, we tweaked a couple of things and focused on what we were doing right. A few days later, the 1st Marine Division charged past al-Kut, bypassing the *Republican Guard Division* of the *Baghdad Division*, which reduced the threat to about five percent effectiveness. Hardly any enemy remained near the city, and . . . most of their weapons had been destroyed. All of a sudden, the morale quickly rose, and the mood throughout the air wing chronicled success again.[37]

Chapter 10
Baghdad Bound

Stalled by Mother Nature

On 25 March, a day after I MEF Marines had successfully crossed the Euphrates River, a massive sandstorm stalled 3d MAW for nearly three days. Sustained winds of 60–70 miles per hour lasted for more than 12 hours and ripped through the heart of the I MEF area of operations, grounding helicopters and many fixed-wing aircraft. The massive storm also created severe conditions for ground forces and U.S. Navy carrier groups in the Persian Gulf. Further worsening flight conditions, opaque layers of dust covered the skies from Baghdad to an-Nasiriyah and blended with black smoke from oil fields deliberately set ablaze by the *Republican Guard*.

Weather conditions began deteriorating a day earlier, halting flight operations in the central and northern area of operations. Air, land, and sea forces had few options other than to wait for the storm to abate.* At its peak, poor visibility essentially paralyzed the 3d MAW helicopter fleet, forcing aircrews to temporarily set down wherever they could find a suitable landing place. Some U.S. Air Force and Coalition fixed-wing crews, however, managed to fly high-altitude bombing missions near Baghdad and interdiction sorties against the Iraqi *6th* and *10th Armored Divisions* in the vicinity of Basrah.

The MWSG-37 weather section provided two daily 3d MAW weather forecasts for General Amos, the tactical air command center battle staff, pilots, and subordinate units. A color-coded system categorizing variations in weather conditions depicted the weather's effect on operations. Ironically, the meteorological forecast on 25 March indicated a "yellow" day for operations, which meant extreme weather was not anticipated.

Despite bad weather and enemy threats, aircrews from HMM-268, under the command of Lieutenant Colonel Jerome E. Driscoll, flew medical evacuation missions in the aftermath of the fighting at an-Nasiriyah. Despite the difficult role, pilots and aircrews ferried wounded Marines to medical facilities south of the town and transport-

*The Naval Research Laboratory in Monterey, California, monitored the situation carefully, and meteorologists remained in constant communication with key commanders via secure electronic mail.

Courtesy of Col Charles J. Quilter II

Marine KC-130s were parked at FOB Joe Foss in Kuwait during the blinding sandstorm that temporarily slowed 3d MAW flights the first week of hostilities.

Photo by SSgt Matthew Hannen, USAF. Defense Imagery 030326-F-MY389-002

Sgts David Irland (left) and Gareth Davies, deployed from the Royal Air Force's 101 Squadron, prepare a replacement engine for a Vickers VC-10 C.1K tanker during a fierce sandstorm that began on 25 March.

Photo courtesy of Marine Light Attack Helicopter Squadron 269

With visibility near zero and ground troops bedded down during the worst days of the sandstorm, aircrews rest at Jalibah airfield.

ed the dead. When the weather worsened beyond safe flight conditions, the CH-46s were temporarily grounded, giving the squadron much-needed rest until they could return to their cockpits.

One morning after only a few hours of sleep, Colonel Driscoll was awakened to hear a mission brief requiring his squadron's immediate assistance. Details were sketchy, but an RCT-5 battalion commander—Lieutenant Colonel Carl E. Mundy III—had requested immediate evacuation for several Marines who were wounded by artillery fire from enemy forces near an-Nasiriyah.

Colonel Driscoll assessed the weather and authorized the mission, selecting Captain Aaron D. Eckerberg as the mission commander. The flight of two CH-46s—flown by Colonel Driscoll and Captain Eckerberg—launched within minutes. After the aircraft landed, Colonel Mundy met the pilots, providing more details about the mission and the circumstances in which a Navy corpsman, Petty Officer Third Class Michael V. Johnson Jr., was killed while tending the wounded in the midst of the battle at an-Nasiriyah. The body was loaded onto Colonel Driscoll's helicopter while a wounded Marine was loaded onto Captain Eckerberg's aircraft. The pilots were soon airborne, heading south toward the nearest hospital facility. A few minutes after liftoff, the helicopters ran into a blinding wall of brown grit. Although they lost communications with each other, both aviators reacted immediately. Captain Eckerberg dropped down to follow the visual markings of a highway, while Driscoll ascended to 4,000 feet to establish aircraft separation, thus avoiding chance of a midair collision.

As Colonel Driscoll flew through the blanket of dust clouds, he relied on his instruments to fight vertigo. With few options, he landed at the closest FARP along Highway 1 to wait out the sandstorm. When a Huey landed to refuel, Johnson's body was transferred to it to expedite the movement of the sailor's body out of the combat zone. It was a fortunate decision because the unplanned stop turned into a two-day ground delay. Captain Eckerberg, on the other hand, successfully navigated the storm and transported the wounded Marine to

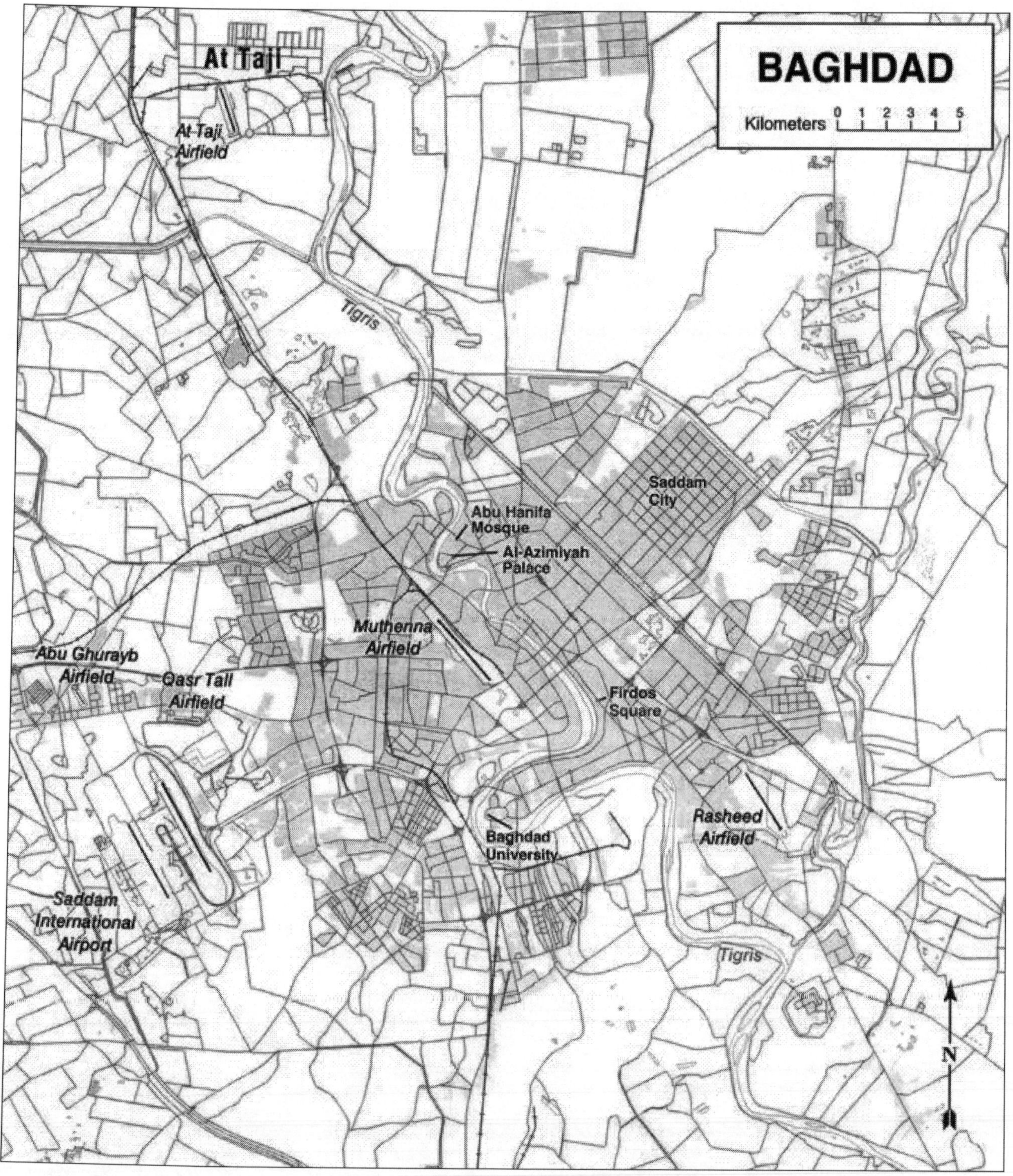

a field hospital where the medical staff provided an encouraging prognosis.[1]

While squadrons and units waited for the weather to clear, commanders continued preparing for the next phase of the war. At Jalibah, I MEF attempted to maintain operations from their headquarters as lights flickered and canvas tent panels flapped violently in the storm. As a precaution, General Mattis assessed the potential risk of the storm's wrath on operations and temporarily transferred forward control of 1st MarDiv headquarters element in Iraq to Camp Commando in Kuwait for about 12 hours, until sustained communications were reestablished.[2]

Critical Joint Assault Support Saves Lives

On the evening of 26 March, a sizeable assault support mission (based on the number of aircraft) originated from FOB Joe Foss in Kuwait. Once the dust storm had abated enough to resume safe flight operations, three C-130Ks from 70 Squadron of the Royal Air Force based in Lyneham, England, arrived and teamed with two KC-130 Hercules from VMGR-452.

The joint mission, flown by pilots Lieutenant Colonel Mark C. Graham, Major William H. Holmes, and Wing Commander John Reid and their crews, was to deliver much-needed logistical resupply at Jalibah airfield and extract wounded Marines from the an-Nasiriyah battlefield. Each British aircraft carried 155mm artillery ammunition, rations, and water, while the two Marine KC-130s hauled JP-8 aviation fuel. One of the Marine aircraft was configured with special racks to transport casualties. Of immediate concern were intelligence reports that Iraqi units with antiaircraft artillery were still active east of Jalibah.

The Hercules air group, dubbed "HAG," depended on recommendations from its operational staff that included VMGR-234 experts Lieutenant Colonels Jerry G. Jamison and Patrick J. Delong and Major Raymond R. Descheneaux. Collectively, they devised a night flight plan to minimize exposure to antiaircraft threats that called for aircrews to use night vision goggles and pilots to employ a mixed route approach into Jalibah airfield.

Major Holmes's aircraft carried a flight crew of three officers and four enlisted Marines, a team of two medical officers and five Navy corpsmen, and a field historian accompanied by a combat photographer. When the aircraft touched down at their destination after a 25-minute flight, the ground crew was unaware of the assault support nature of the mission and struggled initially to accommodate all five aircraft at once. Aircraft commanders and loadmasters had trouble maneuvering their aircraft close enough to the refueling point fuel bladders to unload their cargo because several helicopters were parked on the ramp. Navy Commander Edward W. Hessel had pallets of rations and ammunition unloaded onto the side of the runway because ramp space was unavailable. Once complete, he managed the transfer of 16 wounded Marines, along with 1 patient in critical condition and 1 dead Marine, from ambulances onto Major Holmes's aircraft. With assistance from loadmaster Staff Sergeant Paul J. Morgado, the medical team rigged an oxygen mask for the critically wounded passenger that linked to the aircraft's oxygen system.

By 0200 on 27 March, all aircraft were airborne again, with the British C-130Ks returning directly to their base near Kuwait City, while the two Marine KC-130s flew fast and low toward Joe Foss, using evasive tactics to avoid enemy antiair threats. Once back on the ground at the FOB, the injured were taken off and the body of the Marine killed in action was carried from the aircraft.[3]

Marine KC-130s practice evasive maneuvers to avoid potential enemy antiaircraft fire during a training mission over the Iraqi desert.

Photo by LCpl Andrew Williams. Defense Imagery 030904-M-UW798-012

Photo courtesy of Col Charles J. Quilter II

Shaykh Hantush airstrip was created by Marines from HMM-365 on what was essentially a four-lane highway. The site was planned as a forward refueling and logistics point for the 1st MarDiv before it crossed the Tigris River.

Seizing Shaykh Hantush Highway Strip—The First Time

On 27 March, after three days of gale-force winds and fierce rain, RCT-5 returned to the battlefield during the early morning hours, reenergized and ready to resume their attack toward Baghdad. Colonel Dunford's combat team advanced north along Highway 1 to the next objective—a wide stretch of road called Shaykh Hantush Highway Strip, just 15 miles north of ad-Diwaniyah. Despite its austere surroundings, it was an important piece of terrain because it provided access to the east-west corridor that General Mattis planned to use as he approached Baghdad from the eastern flank. However, it soon became an important air operations hub located just 100 miles south of Baghdad.

After the area was secured, the air-ground team planned to transform the highway into a FARP, referred to as "QualCom," which would be one of the final resupply points for ground forces before crossing the Tigris River and moving to Baghdad.[4] Reliance on Marine aviation had grown steadily since the beginning of combat operations, and the makeshift Shaykh Hantush airstrip would become an essential logistical link in that lifeline. KC-130s would deliver their cargoes of food, ammunition, and fuel in a major replenishment effort to support 1st MarDiv.

The division had endured days of a "logistics lite" policy—the rapid movement of troops without extra equipment or supplies—after they passed through an-Nasiriyah.* Until this point, major resupply missions inside Iraq had been accomplished through ground transportation. With each day of the drive to Baghdad, however, the tactical

*The "logistics lite" philosophy is similar to the commercial materials management term for when supplies are delivered on a "just in time" basis.

convoys became increasingly vulnerable to enemy attacks as the supply lines stretched. The Shaykh Hantush airstrip eliminated the need for supply by long-range ground convoys.

Visibility was still poor, but RCT-5 moved north, focused on securing the airfield despite encountering scattered pockets of enemy fire by irregular forces on the sides of the highway. As the combat team neared its objective, enemy contact intensified, particularly with several hundred fedayeen

"Logistics Lite" Philosophy

Senior commanders understood that ground troops saddled with excess equipment and supplies would move more slowly, yet dependence on ground logistical resupply efforts placed tremendous risks on ground convoys fed by long supply lines on Iraqi roads. Thus, the 1st FSSG commander, Brigadier General Edward Usher, and his deputy commander, Colonel John L. Sweeney Jr., studied in-depth the spectrum of logistical problems that would likely be encountered in supplying fast-moving ground and aviation combat elements over the 400 miles from northern Kuwait to Baghdad.

Their resulting logistics strategy aligned with General Mattis's philosophy to fight lean. Unnecessary equipment and supplies would not be carried by assault troops. They developed a "logistics lite" concept that eliminated physical burdens from the rapid movement to Baghdad. Their plan envisioned the Marine air-ground team eschewing long, vulnerable ground supply lines and instead relying on short-notice resupply via air when needed. This was a new concept for the Marine Corps that challenged air-ground logistical planners unlike any previous campaign. They had to plan methods to deliver supplies to remote airstrips and forward sites deep inside Iraq. This strategy relied on Marines using only organic supplies—those they carried with them—after they crossed the Euphrates River just north of an-Nasiriyah.[5] Small combat service support companies would support regimental combat teams to satisfy their logistical needs. Numerous I MEF situation reports repeatedly cited General Conway's satisfaction with the new approach of logistical support, especially the fuel line hose stretching from Kuwait to Jalibah that supplied fuel throughout the campaign.[6]

General Usher's logistical planning also incorporated contingencies. The far-reaching support mission for I MEF was delegated to the Army's 337th Theater Support Command, yet its representation during 2002 predeployment site surveys alerted planners to anticipate a potential gap in providing basic sustainment items, such as food, water, fuel, and ammunition, to frontline Marines.[7] Fortunately, when deficiencies in the Army's system surfaced, Marines could rely on their combat service support companies to provide the basic supplies. An exception that was never fully resolved was the long-standing, systemic spare parts shortages for tactical equipment. Many modes of transportation were used to get needed supplies to troops deep inside enemy territory, including Marine transport and third-party national contractors. General Usher characterized the Marines' efforts this way: "It was a matter of brute force logistics. It wasn't pretty, it was not elegant. It was just sheer adrenaline."[8]

On board the USS Kearsarge *(LHD 3), Navy aircraft handler PO3 Curtis Turner directs a Super Stallion from HMH-464 for a combat resupply mission into southern Iraq while a second helicopter awaits liftoff clearance in the background.*

Photo by PO3 Angel Roman-Otero, USN. Defense Imagery 030325-N-CF941-032

who had fled from ad-Diwaniyah to join fighting closer to Baghdad. Although the enemy retreated, the Marines encountered sporadic firefights during the seizure of the airstrip. Marines engaged paramilitary forces armed with a mix of RPGs, mortars, and heavy machine guns near the cloverleaf intersection of Highways 1 and 17 just east of ad-Diwaniyah. But Marine Corps tanks, artillery, and tactical aircraft counterattacked, destroying enemy tanks and antiaircraft guns.[9] By midmorning, the Iraqis were defeated and had either retreated or been killed during the fight. The Marines seized the airfield within four hours, yet no one suspected that, by late afternoon, Colonel Dunford would be forced to halt his offensive again, not due to weather but by a command decision.[10] He would, however, return three days later to secure the same airfield for 3d MAW operations.[11]

By now, flying conditions had improved throughout the 3d MAW airspace, and it looked as if the wing would return to a busy flight schedule. The dust storm was followed by a heavy soaking rain that transformed the landscape into fields of mud and colored the sky with an artist's pallet of deep orange hues. Fixed-wing aircraft focused on shaping operations directed against *Baghdad's Republican Guard Division* and resumed interdiction missions near Basrah. Despite the muck, the wing returned to a normal pace with a full schedule, logging 361 missions. However, the tempo of air operations soon slowed again.[12]

Relying on years of training and mobile communications experience, SSgt Charles C. Robinett controlled hundreds of helicopter and fixed-wing aircraft at several FARPs while assigned as a tactical air controller attached to MWSS-371.

Photo courtesy of Cpl Nathan A. Wachter

Operational Pause: Halting for Bullets, Beans, and Band-Aids

Despite the speedy and intensive effort to secure Hantush airstrip, RCT-5 received a surprising order from the commander of Coalition Forces Land Component Command (CFLCC), Army Lieutenant General David D. McKiernan, to withdraw from the airfield and focus on rear area security. The joint staff referred to the withdrawal as "an operational pause." Although General Mattis disagreed with the decision, he executed it. On 27 March, the 1st MarDiv commander forwarded the order to his subordinate commanders, directing them to pause in their attack north and to look to the rear to concentrate on securing lines of communications. General Mattis had difficulty issuing a withdrawal order to Colonel Dunford, who had just captured the airfield in merely four hours and had lost men during the fight. Despite this, 1st MarDiv moved into a defensive role and awaited further guidance.[13] Questions arose regarding another delay en route to Baghdad. Additional inquiry by senior Marine commanders confirmed that the pause was a decision by General McKiernan because of his concern, expressed earlier by the U.S. Army V Corps, that it had exhausted most of its supplies during the sandstorm. The consensus among the CFLCC staff was that a major resupply was needed before proceeding to Baghdad. By evening, I MEF Marines began retracing their route moving south. They refrained from calling the movement a "withdrawal" and, rather than dwelling on disagreement regarding the pause, they established positions near the highway cloverleaf to focus on securing the immediate area.

During the Lull

Although the operational pause ordered by the CFLCC temporarily slowed the march to Baghdad, the Marines were not idle. During this period, ground forces focused on security patrols, cleaning weapons, and mental preparation for the attack on Baghdad while combat service support Marines resupplied units for the next offensive phase. General Mattis directed his commanders to focus on destroying the fedayeen presence in nearby ad-Diwaniyah because intelligence reports indicated that the city had become a base for displaced terrorists who desperately wanted to regroup against Coalition forces. The local population was powerless against the gangs of violent fighters clothed in black who controlled their streets. The terrorists continued their tactics of feigning surrender under white flags when close air support aircraft flew overhead, only to resume small-arms fire moments later.[14]

General Amos used the pause to shape the battlefield ahead of the ground forces with aggressive air strikes against enemy artillery, tanks, and ammunition storage sites deep inside the I MEF zone.[15] Because the ground war had started earlier than planned and poor weather had grounded many flights, shaping had been minimal. Fixed-wing aircraft now focused on deep interdiction efforts against the *Baghdad Republican Guard Division* in the vicinity of al-Kut and against the *6th* and *10th Divisions* located near al-Amarah and Basrah to protect I MEF's eastern flank. The Hornet squadrons were instrumental in controlling and coordinating strikes and aerial reconnaissance missions.[16] By 29 March, 3d MAW had flown 341 missions, which indicated that air operations were back to normal despite unplanned interruptions and the operational pause.

Hail to a Hasty FARP

During the division's advance, the air wing's five flying groups depended on its wing element to keep pace—MACG-38 for air control and MWSG-37 for aviation ground support. On 28 March, these units demonstrated the epitome of teamwork and gained unsolicited attention when enemy forces ambushed a 250-vehicle convoy from MWSS-371 on Highway 7 near Ash Shatrah. The squadron had spent the last several days in an-Nasiriyah and was leapfrogging northward to set up a new FARP at Qalat Sikar, later referred to as Fenway. Although the squadron requested aerial escort as it departed an-Nasiriyah, no air assets were immediately available. Regardless, the convoy departed in the midafternoon and soon found itself amid a winding string of Coalition vehicles. As the convoy stretched over miles, it morphed into a massive traffic jam, creating a prime opportunity for an enemy ambush.[17]

Staff Sergeant Charles C. Robinett, a Marine air traffic control team leader attached to MWSS-371, was riding in the convoy when enemy fire erupted. Realizing the convoy desperately needed airpower assistance, he submitted two joint tactical air support requests to the airborne direct air support center: one was to provide immediate close air support to attack enemy positions, and the second was for an armed escort as the convoy continued its journey north. While Cobras did not arrive on-station when requested, random enemy fire stopped. As the sun began to set and visibility slowly began to fade, however, the convoy was attacked again. Corporal Nathan A. Wachter, a fellow tactical air controller, was not surprised because he knew that the enemy liked to attack at dusk when night-vision goggles were least effective. Having just completed a 2d MAW tour in Afghanistan during OEF, Corporal Wachter had become adept at combat operations, yet this evening proved to be one he would never forget.

Relieved when he saw four armed AH-1W Cobras circle overhead, Sergeant Robinett nevertheless grew anxious when the aviators relayed that their time on-station was minimal because of low fuel. Even more disturbing was the direct air support center (airborne) broadcasting the convoy's location by using its 10-digit grid coordinate* as the same area where three of the most-wanted Iraqi officials were reportedly located. Given limited options, Corporal Wachter believed their situation was dire and worsening with each passing minute. From his perspective, he worried the convoy would sustain casualties and recalled, "We still had a lingering firefight at an-Nasiriyah to the south; to the north, three wanted Iraqi officials were suspected near our location, which threatened our friendly position even more. I thought, how do we want our bacon fried?"[18]

Sergeant Robinett combined innovation with sheer determination to defend the convoy and devised an unconventional plan that was readily approved by the FARP officer-in-charge. After coordinating briefly with the airborne direct air support center and gaining pilot concurrence, MWSS-371 Marines prepared for a hasty refueling mission on the dirt road in front of the convoy to keep the Cobras on-station. Without body armor or concern for his own safety, Sergeant Robinett

*A 10-digit coordinate provided pinpoint accuracy to a location within 1 meter, which was critical when controlling close air support missions.

jumped from his vehicle with only a manpack radio and cleared an area as a makeshift landing zone, where he began controlling the refueling operation. Communicating under fire, the sergeant directed the aviators to the landing site while Corporal Wachter provided security. Two Cobras landed and hot refueled, while the other pair of Cobras circled overhead. Sergeant Robinett and Corporal Wachter ran the hasty operation as the convoy moved forward in the dark of night, coordinating a total of 12 close air support and escort sorties flown by Hornets and Cobras.[19]

Despite a series of firefights near Ash Shatra, the convoy reached Qalat Sikar and established a communications node for the FARP by dawn the next day. One Marine was killed during the attack and another wounded, but Marines responded and minimized losses or injuries.[20] Named Fenway, the FARP became even more crucial when ground convoys were halted temporarily along Highway 7. Marines had encountered stiff resistance from paramilitary fighters along this route, which endangered ground transportation for almost a week.[21]

A Leadership Meeting

On 28 March, Army Generals McKiernan and Wallace and Marine General Conway convened a senior leadership council at Jalibah airfield to discuss the upcoming Army and Marine Corps coordinated attack into Baghdad. The Army's focus was on securing the rear areas, completing the resupply effort, engaging the *Republican Guard* and Iraqi paramilitary forces as needed, and seizing the capital. After the meeting, General McKiernan issued Fragmentary Order 040-03, which outlined details from the discussion.

General Conway focused on the final leg of the journey and requested a revised cross-country mobility study from the Marine Corps Intelligence Activity. The study outlined options for the best route from al-Kut to Baghdad without using main roads. The intelligence team reviewed recent digital imagery and considered historical late winter and early spring weather conditions before recommending a direct approach via Highway 6. Another option that had previously been considered was called "Hook Two," which entailed an east-

LtCol Wendy A. Smith, commanding officer of MALS-13, points out an AV-8B Harrier Litening II targeting pod.

Photo courtesy of Col Charles J. Quilter II

ward swing south of the city, then an approach into it from the northern perimeter. The intent was to avoid main roads, such as Highway 6. As briefers outlined the limited options, the I MEF commander listened intently, studied the options, and probed for information. After about five minutes, General Conway thanked senior liaison officer, Lieutenant Colonel Brian L. Sulc, for the briefing. All who attended the briefing clearly understood that 1st MarDiv would advance into Baghdad by way of Highway 6.[22]

Jump-Starting the March to Baghdad with Harriers

By 28 March, the 3d MAW's flight schedule doubled to make up for the slower tempo caused by the weather and operational pause. General Robling, assistant wing commander, along with wing Sergeant Major Estrada, were among the first from the command element to move forward, flying to Jalibah airfield. Just four hours before VMA-311 launched its first flight of the morning on 31 March from the *Bonhomme Richard*, the squadron's commanding officer, Lieutenant Colonel Michael K. Hile, received a late change to the air tasking order. He was assigned a special high-interest mission against eight priority targets to support a predawn I MEF raid into Ash Shatrah. The targets consisted of a cluster of buildings within the city limits near Baath Party headquarters. The mission presented major concerns of fratricide and collateral damage, so Harriers were chosen because of their proven ability to deliver a 500-pound bomb with pinpoint accuracy.[23]

Colonel Hile, flying as mission commander with Captain Duncan as his wingman, led four sections of Harriers—two from VMA-311 and two from sister squadron VMA-211. Each section was assigned two targets in proximity to one another with the intent of striking all eight targets within eight minutes. Colonel Hile devised a simple plan that reduced conflict of elements by altitude with a coordinated attack from each of the four cardinal directions to minimize potential overlap or confusion. Before the launch, he realized that three of the eight targets were on the CentCom collateral damage list and required joint command approval before striking them. Final clearance to engage all eight targets was not received until the aircraft were airborne, which meant a formal request to drop ordnance was passed through the Marine Corps' command-and-control agencies, specifically, the direct air support center as the first tier, followed by a transfer—or hand-off—to the tactical air operations center as the final controlling agency.

Check-in protocols were followed precisely without delay and the timed strike went as planned. Bomb damage assessments confirmed six of the eight targets were successfully hit without collateral damage. Mission success was attributed to pilot skill; precision-guided munitions; the Litening II targeting pod; and the ability to rapidly plan, approve, and execute missions among controllers within the Corps' command-and-control agencies.[24]

On 30 March, Harrier pilots Captains Guy G. Berry and Gregory Warrington of VMFA-211 received unexpected praise from embedded journalist and retired Marine Oliver L. North.* The Huey that North had been riding in was forced down because of engine trouble in the vicinity of an-Numaniyah, about 60 miles south of Baghdad. The helicopter landed safely but immediately began receiving enemy artillery and mortar fire. The crew anticipated enemy infantry would soon move toward their position and requested close air support. Colonel Savarese, MAG-13's commanding officer, recalled the situation:

> We had a section of Harriers airborne at the time. The pilots had just finished another mission when they received a request for close air support. They targeted enemy positions closest to the Huey helicopter and provided delaying action against enemy infantry moving west. Once cleared, Marine rescue helicopters landed without fire and ferried the aircrew and passengers on board to safety. It wasn't until afterward we found out that Oliver North was one of the guys in that helicopter. Our squadron was pretty proud.[25]

Hercules Shuttle

On 30 March, General Conway met with General Mattis at Jalibah to discuss the next steps in jump-starting the ground scheme of maneuver. They confirmed their plan to cross the Tigris River, iso-

*LtCol North, a key figure in the Reagan administration's Iran-Contra Affair while on active duty, was viewed as a friend of the Marine Corps.

Photo courtesy of Col Charles J. Quilter II

Although MajGens James N. Mattis (left) and James F. Amos communicated at least once daily via advanced telecommunications media, they chose to meet in person to discuss the air-ground scheme of maneuver as 1st MarDiv approached Baghdad.

late al-Kut, anticipate release of weapons of mass destruction, and begin encircling Baghdad. They further decided to employ Task Force Tarawa during the Baghdad offensive by expanding its battlespace.[26]

The next day, General Amos finalized operational details with General Mattis before the Marines advanced the last 100 miles into the capital city where Saddam's defenses were expected to be concentrated. Amos had promised Mattis months before the start of OIF that he would do everything he could as the wing commander to support 1st MarDiv. Undoubtedly, the two generals shared a unique bond, which influenced the way the air-ground team executed the air campaign, so General Amos arranged for a personal meeting with the division commander before the final push.[27]

General Mattis, What Do You Need?

Believing the daily video conference or satellite phone call would not suffice, General Amos traveled six hours round-trip to meet with General Mattis on 31 March, flying barely 100 feet off the ground in a CH-46 Sea Knight to avoid surface-to-air threats. On the eve of the final move into Baghdad, the generals met outdoors and discussed plans over maps laid out on the sand. General Mattis described his plan for conducting a feint northward with RCT-5 moving east and consolidating with RCT-1 and RCT-7 at Highway 6 before attacking toward Baghdad.

The meeting did not take more than a few hours, yet it confirmed the next steps. General Amos promised to deliver whatever General Mattis needed most, which was a shuttle of KC-130s filled with food, supplies, and ammunition for the Marines, before they made the last leg of the journey north toward Baghdad.[28]

Before returning to al-Jaber Air Base, General Amos took a side trip to see where 3d MAW would stage one of the most important forward air operations of the war—Shaykh Hantush Highway airstrip, a portion of Highway 1. The need to resupply 1st MarDiv before the Baghdad offensive began meant KC-130s had to begin delivering supplies and cargo in less than 24 hours. General Amos wanted to see exactly where the Hercules crews would be sent. As his helicopter circled over a large grove of date palm trees and prepared to land, he saw a welcome sight: Colonel Dunford's RCT-5 vehicles along the highway. Later, Amos learned that dozens of enemy fighters had been hiding in the trees as he flew in, yet surprisingly, none of them took a shot at his aircraft.

During his tour of the area, the general assessed the 8,000-foot runway section of highway and asked fellow aviator, Colonel Quilter, who had accompanied him on the trip, for his opinion about the feasibility of landing KC-130s on the airstrip. The colonel, a retired reserve officer on a leave of absence from a civilian airline, kicked the sand near the runway edge with the heel of his boot several times and replied that he could land a FedEx airplane on the runway. The general nodded and declared the airfield fit for operations; within 24 hours, Marines welcomed the roar of KC-130s.[29]

Shaykh Hantush airstrip was not an ideal site for air operations, but the divided superhighway sufficed as a temporary runway for one of 3d MAW's most important missions. General Amos realized that neither Hercules aircrews nor aircraft maintenance personnel would enjoy the austere, sandy conditions, but the airfield was more suitable than other options. Once Amos returned to al-Jaber, he briefed his staff about the upcoming mission. RCT-5 returned to Shaykh Hantush and seized the area within a few hours, this time meeting minimal resistance. Despite indirect fires, RCT-5 engaged the enemy, capturing 15 prisoners and destroying 12 antiaircraft artillery pieces. MWSG-37 engineers followed closely on the heels of RCT-5 and prepared the airfield for operations by installing bladders to support a critical fuel farm. The capture of Hantush airstrip was a significant victory that not only provided a major air operations hub, it also reinforced an enemy perception that the Marines would continue their march to Baghdad up Highway 1.

By late evening on 31 March, a steady flow of deliveries began at the airstrip as a rotation of KC-130s landed and unloaded much-needed supplies and fuel for 1st MarDiv. Aircraft had been positioned at FOB Joe Foss in Kuwait, which reduced flying time and expedited the synchronized shuttle operation. Hantush airstrip's significance could not be underestimated. It was the first Hercules-capable hard-surface airfield north of the Euphrates River, and its proximity to the frontlines facilitated casualty evacuations to medical facilities in rear areas.

Colonel John A. Toolan, RCT-1's commander, praised 3d MAW, particularly Colonel Anderson's MWSG-37 Marines:

> The wing support personnel provided first-class logistical support. At one time, we didn't have any meals ready-to-eat, or MREs; we weren't starving, but we had little chow left. During that opening night of the shuttle, we watched the KC-130s flying in on the highway that our unit just crossed. It was pretty cool. It was a motivational thing.[30]

Crossing the Tigris River

Employing Aviation to Bypass and Isolate al-Kut

On 1 April, the operational pause ended almost three days of slowed operations. Although the decision frustrated Marine Generals Conway, Amos, and Mattis, it became a footnote in command chronologies as the air-ground team used

Map courtesy of 1st Marine Division

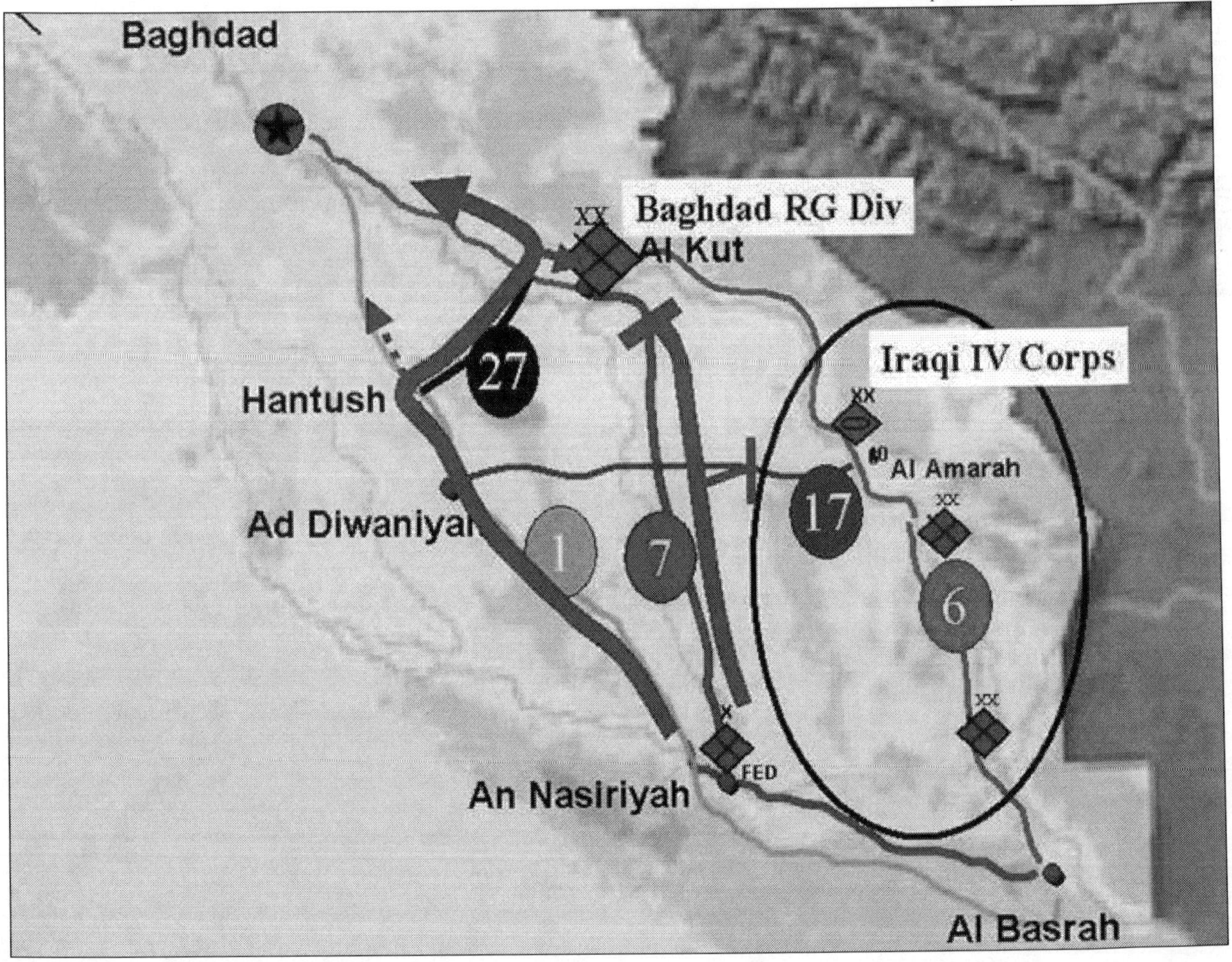

Photo by LCpl Alicia M. Anderson.
Defense Imagery 030225-M-5607A-010

Assigned to VMU-1, Cpl John Rocha monitors the GCS-2000 Ground Control Station that controls UAVs. MACG-38 oversaw the Maine Corps' only UAV, which proved invaluable in providing ground commanders with real-time information of enemy activity.

the period to prepare for regaining momentum for the Baghdad offensive.[31] Still, approximately 100 miles south of the capital, 1st MarDiv faced another major waterway—the Tigris River. Unlike the Euphrates, this bridge crossing presented less vulnerability for ground forces. RCT-1 protected the main division effort from their position south of al-Kut on Highway 7 and served as a fixing force against the *Baghdad Republican Guard Division*. RCT-5 crossed its line of departure and headed east on Route 27 and crossed the Tigris, establishing a blocking position at Highway 6 between Baghdad and al-Kut. Meanwhile, RCT-7 continued attacking north on Highway 1, past ad-Diwaniyah, where it captured an enormous 40-bunker ammunition supply point of the *Saddam Fedayeen* and *al-Quds* militias.[32] After RCT-7 crossed the river, it joined the division and relieved elements of RCT-5 near the town and airfield of an-Numaniyah.

Searching for enemy activity ahead of RCT-1, VMU-1 traveled by convoy from Jalibah to Qalat Sikar up Highway 7. By evening, the squadron's UAVs began flying from a fresh FARP, coordinating missions with its sister squadron VMU-2, which remained at Hantush airstrip. The squadrons provided imagery to the ground forces that aided in quickly securing an-Numaniyah Airfield and al-Kut.

Sections of fixed-wing aircraft continued shaping the vicinity of al-Kut as early as 23 March, providing constant aerial cover and focusing on *Republican Guard* forces well ahead of 1st MarDiv. With 3d MAW F/A-18D Hornets as their platforms, FAC (airborne) aircrews coordinated strikes and flew armed reconnaissance missions at targets around the town, once again employing aviation as a separate maneuver element.

By the evening of 2 April, 1st MarDiv surrounded al-Kut and was in position to destroy the *Republican Guard* with the help of 3d MAW airpower and 11th Marines artillery. Damage assessments showed that a significant portion of Iraqi armor and artillery in the area was either destroyed or abandoned. The Marines, despite physical obstacles and treacherous terrain viewed as untraversable, had succeeded in diverting attention from the Army's main effort and drawing enemy troops from the capital city to engage 1st MarDiv.

The Iraqi military began unraveling outside Baghdad. With only a small number of fedayeen and paramilitary fighting units intact, the regime had lost communications, control, and access to move openly from Baghdad to Basrah.[33] As the Army and 1st MarDiv tightened their stranglehold on Baghdad, they focused on two mechanisms. One was to block enemy forces from moving from the south and east to the city, and the other was to draw Saddam loyalists and paramilitary fighters out of urban areas away from innocent Iraqis. By preventing reinforcements from entering Baghdad, the Coalition would further degrade the enemy's ability to communicate and their willingness to fight.

Rescuing an American Prisoner of War

As 1st MarDiv approached the outer city limits of Baghdad on 1 April, Coalition forces engaged in the dramatic rescue inside an Iraqi hospital of Private First Class Jessica Lynch, USA. Her nine-day captivity captured the international spotlight after her Army unit, the 507th Maintenance Company, had been ambushed in an-Nasiriyah. The rescue was a joint operation involving each of the four main U.S. Services—Air Force AC-130 gunships and A-10 Warthogs; Marine Corps AV-8B Harriers and helicopters; elite units from the Army's Special Forces and Rangers; and the Navy's SEAL forces.

Official U.S. Marine Corps photo

BGen Richard F. Natonski (center, sitting with glasses in hand), Task Force Tarawa's commanding general, plans for the joint rescue operation with U.S. Army special operation forces.

Designated Task Force 20, the mission resulted not only in the rescue of Private Lynch, but outwardly demonstrated the U.S. military's commitment to recover its captured and fallen troops.[34] Understandably, Task Force Tarawa was eager to participate in this particular mission, especially since it involved a soldier from the unit the Marines had defended a week earlier. Task Force Tarawa teamed with Task Force 20 to develop a rescue plan based on sketchy information provided by an Iraqi informant about American captives and their location. Task Force 20 colocated its command post with Task Force Tarawa's combat operations center to craft a plan for a Marine Corps helicopter lift, ground troops for landing zone control and security, and an Army Special Forces team inside the hospital for the rescue. Task Force Tarawa would provide countersniper support, deception attacks, and also a reaction force to extract Task Force 20 in the event that significant opposition was encountered.[35]

The 3d MAW planners were extensively involved in the operation because the complex plan required Marine Corps aircraft for assault and close air support along with an EA-6B Prowler to suppress enemy air defenses. The wing provided assault support lift helicopters for the Special Forces insertion along with close air support and armed reconnaissance using the Litening II–equipped Harriers.[36] The commander of MAG-16, Colonel Stuart L. Knoll, took the lead mission planning role and assigned Lieutenant Colonel Gregg A. Sturdevant, the commanding officer of HMM-165, as the assault flight leader along with Lieutenant Colonel Daniel C. Deamon, the HMH-465 squadron commander.[37] MAG-13's commanding officer, Colonel Savarese, tasked a division of Harriers to assist. VMA-211, commanded by Lieutenant Colonel Kevin S. Vest, provided the AV-8Bs for close air support and armed reconnaissance over the hospital complex.

Under cover of darkness and with guidance from aircraft with global positioning systems and night-vision devices, HMM-165 and HMH-465 launched from Tallil Air Base, about 12 miles west of an-Nasiriyah. The aircrews flew a mix of 10 CH-46 Sea Knights and CH-53 Stallions and successfully inserted 288 Army Rangers in two waves; the CH-46s arrived first, followed by the CH-53s. Once on the ground, the Rangers quickly established perimeter security and then stormed the hospital, searching for American prisoners.[38] Showcasing the Litening II pod's capabilities, VMA-211 collected digital imagery and provided a real-time stream of the rescue operation in progress for the tactical air command center battle staff.[39] The squadron's intelligence staff assisted the pilots in gathering vital digital imagery and footage from the pod sensor and storing it in an online central collection repository. In turn, aircrews and squadron intelligence personnel quickly forwarded video imagery from mission collections for the air-ground team that pinpointed potential enemy target locations and provided a more detailed picture of the enemy disposition. This timely distribution and sharing of data among joint commands often triggered national asset collection requests for additional details and analyses. As a byproduct of this innovation, the intelligence section of VMA-211, led by First Lieutenant Alan L. Ramsey, developed new procedures for intelligence gathering in support of combat missions.[40]

The joint mission resulted in Lynch's rescue along with the recovery of nine bodies of members of her unit (the remaining members were discovered a few weeks later). The success of this mission generated a much-needed morale boost among those Marines who had endured losses near an-Nasiriyah less than two weeks earlier.[41]

In a daring rescue mission conducted by U.S. Special Forces and Task Force Tarawa, PFC Jessica D. Lynch, USA, 507th Ordnance Maintenance Company, is returned to military custody after enemy captivity in an an-Nasiriyah hospital.

Defense Imagery 030402-D-0000X-004

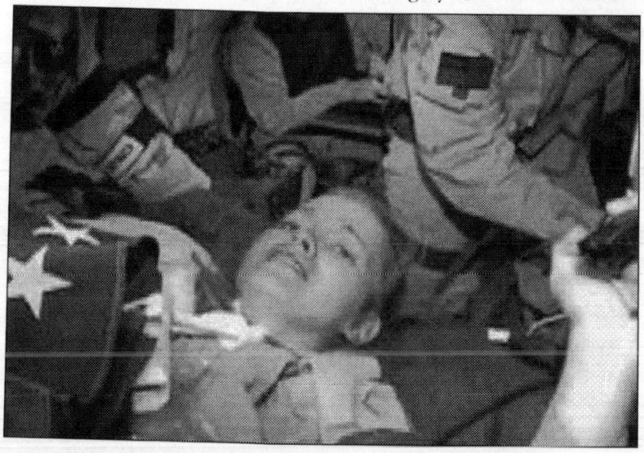

Patrolling Over Baghdad

From 19 March to 5 April, the air wing's statistics told the story each day and confirmed that Marine aviation, with few exceptions, had supported Marines on the ground. While the Air Force conducted strategic bombing missions over Baghdad during the initial days of OIF, Marines flew coordinated attacks in simultaneous air-ground operations with 1st MarDiv. The air tasking order, or master flight schedule, reflected that the wing flew most of its missions in direct support of I MEF until 5 April, when the Army's V Corps and 1st MarDiv circled the outskirts of Baghdad and began requesting Marine air support for joint missions.

Believing that Baghdad was Saddam's center of gravity, CentCom projected that the dictator's most brutal capabilities—the *Republican Guard*, the *Special Republican Guard*, and chemical weapons of mass destruction—would confront Coalition forces there. It was expected that Saddam's conventional forces would establish defined layers of defense around the city to protect the inner core of power—the heart of command and control, or the "Red Zone"—and its geographical boundaries that extended to Karbala on the western flank and al-Kut on the eastern.

Senior-level joint planners representing the Air Force, the Army's V Corps, UK forces, and I MEF differed in their approaches of how to most effectively secure Baghdad. They had long debated this topic even after combat operations had begun. Options ranged from the Army's concept of systematically conducting a series of raids in and out of the city to British lessons learned from skirmishes in Northern Ireland that favored more ground forces augmented with pinpoint snipers and shying away from heavy mechanized armor. General Mattis expressed a more traditional infantry approach coupled with a dependence on combined arms that easily translated into the Marine air-ground task force model.[42] The Marine Corps' Warfighting Laboratory at Marine Corps Base Quantico, Virginia, had spearheaded several war game scenarios and exercises that focused on urban combat that began as early as 1999. At Quantico, the crossroads of the Corps' officer and staff noncommissioned officer training as well as home

Photo by LCpl Andrew P. Roufs. Defense Imagery 030403-M-DC117-102

Marine AH-1Ws provide close air support for elements of the 1st LAR, 1st MarDiv, during an ambush by Iraqi forces in northern Iraq.

of Marine Corps University's top-level educational residence and senior commanders' courses, urban operations were a familiar topic.[43]

On 3 April, General McKiernan assessed the situation and finalized the Baghdad attack plan, issuing Fragmentary Order 124 that outlined the geographic boundaries between I MEF and V Corps. The military order clearly divided territory between the two commands at the Tigris River (with the Marines on the east) and described a coordinated attack. Coalition forces moved around the outskirts of Baghdad much sooner than the 55 days originally projected by 3d MAW planners.[44] As 1st MarDiv headed north, the air wing began supporting an increasing number of joint missions. Overall, the number of missions flown in direct support of Marines steadily declined as the sortie rate increased for direct support—up to 25–30 percent—for the Coalition FLCC offensive. Aircrews established a 24-hour presence overhead, maintaining a FAC (airborne) to direct close air support attack aircraft.

On 5 April, the wing's assignments included 14 reconnaissance missions in support of I MEF collection requests; strikes against two military camps near al-Amarah and Batra with Marine Corps, Navy, and UK aircraft; and a coordinated attack with Air Force B-52 high-altitude bombers near al-Amarah. As ground forces circled Baghdad for the final offensive, the wing continued providing Marine aviation overhead until the city was seized.[45] Eventually, FACs declared air supremacy over Iraq, and the first Coalition aircraft landed at Baghdad International Airport, signifying major air campaign milestones.[46]

Meanwhile, Marines and Army ground forces attacked along the steep and muddy banks of the Diyala River that emptied into the Tigris River, gaining more territory around the capital. The Diyala River crossing was not as easy as originally briefed, but it was a critical and unavoidable path to Baghdad. Regardless of the physical obstacles, the Marines expanded a bridgehead over the river and captured the Rasheed military airport on the eastern edge

Photo by LCpl Andrew P. Roufs. Defense Imagery 030402-M-DC117-022

Armed with AGM-114 Hellfire missiles, Marine Cobras provide close air support as light armored vehicles with the 1st MarDiv move into position during a firefight on the outskirts of Baghdad.

of the city. RCT-5 advanced to Baghdad from the north while RCT-7 became the lead unit, advancing from the south with RCT-1 on its flank.[47] The wing aircrews focused on the *al Nida Republican Guard Division* west of Baghdad and the *Medina Republican Guard Division* to the south, along the Tigris. As RCT-5 moved south, attack aircraft ensured that enemy forces did not follow the unit's path toward the city or exit it to build a defensive perimeter.[48] Additional 3d MAW and joint air strike missions were directed at the Rasheed military airport complex to secure it as well.[49]

Salman Pak East Airfield

By 8 April, another challenge arose that required immediate assistance from 3d MAW. General Mattis reported that the division had just a one-day supply of rations and ammunition remaining as the division focused on joining V Corps for a final southern assault toward Baghdad.[50] As the division consolidated forces and advanced northward along Highway 6 to begin the offensive with the Army's 3d Infantry Division, the air wing prepared to establish another FARP. General Amos had studied maps in anticipation that the wing might need another airstrip closer to Baghdad, particularly to support helicopter operations nearer to the fight, so he selected Salman Pak. This was a suitable airfield in a small town about 20 miles southeast of Baghdad, not far off Highway 6. Ironically, area citizens reported that it had been a training camp for Iraqi special forces, where a Boeing 707 fuselage was allegedly used for aerial hijack scenario training.[51] Although it had a longer runway than an alternate option, there was no intelligence regarding its condition or the possibility of enemy threat. The air wing needed help securing the runway, and General Mattis quickly obliged by tasking a light armored reconaissance unit to secure the runway, which rid the airstrip and adjacent areas of visible enemy threats.

Supported by Cobras overhead, RCT-5 once

again led the advance and, with some casualties, successfully secured the airfield. Lieutenant Colonel Chris A. Lamson's MWSS-372, along with engineers from a Navy Seabee detachment, then cleared obstacles to open the 9,000-foot-long runway. Once 3d MAW gained control of the runway, General Amos called Colonel Milstead, the MAG-29 commander, and conveyed his plan: "The attack on Baghdad has started and RCT-5 is running out of artillery shells. They need food. I want you to take your MAG-29 flag and a half a dozen of your attack helicopters and some logistics—I want you to establish a presence. I want to have an aviation command element on the ground at Salman Pak today!"

As General Amos recalled,

> Colonel Milstead positioned a mix helicopters up there—at first, two Hueys, four Cobras, four CH-53s, and two CH-46s, along with additional Hueys and Cobras later.[52] Assigned to RCT-5, Marine Wing Support Squadron 371 set-up fuel bladders, and by nightfall, KC-130s began deliveries of ammunition and supplies. I had one of the largest KC-130 deployments ever recorded in one location, so we filled KC-130s front to back with the meals-ready-to-eat [MREs] and ammunition for these guys. I had no idea where the food would come from, but I just assumed the logisticians got them from a supply point in Kuwait. Later I learned my guys retrieved the meals from nearby 3d MAW stockpiles and wherever else they could find more stashes to give to the Marines who were headed into Baghdad.[53]

The KC-130s flew all night onto a very narrow runway, probably 75 feet wide and 8,000–9,000 feet long. The runway was in fairly good shape because the engineers cleaned it off, but most runways are about 150–175 feet wide. Since there was no room to turn around, pilots would land, reverse the engines, back up, and go forward until the aircraft was turned in the opposite direction for immediate takeoff. "It was very, very varsity flying," noted Amos. The next morning during the morning video conference call, General Mattis repeatedly praised 3d MAW.[54]

Once secured, Salman Pak became an active

A KC-130 Hercules from VMGR-234 delivers cargo at Salman Pak. The airfield became a logistical hub of activity with a steady flow of aircraft delivering food, supplies, and ammunition to the 1st MarDiv just days before the Baghdad attack.

Photo courtesy of Col Charles J. Quilter II

hub of air operations for the final leg of the Baghdad offensive. One of the first tenants to arrive was VMU-2 with its RQ-2B Pioneer UAVs. The squadron had been leapfrogging from one FARP to another, supporting I MEF throughout combat operations. Once settled at Salman Pak, flights resumed and operators focused on collecting imagery within a few miles of Baghdad, providing full coverage of the city to detect enemy movement. VMU-2 flew 133 combat hours and 32 sorties in support of 1st MarDiv. When no viable targets remained because they had either been destroyed or had deserted, the squadron packed up again on 20 April and headed south toward al-Kut to operate from Blair Field.[55] Despite the rapid pace, squadron morale remained high and was boosted even more when Marines received their first mail since crossing the border from Kuwait.

Meanwhile, VMU-1 rotated between units, flying from Qalat Sikar the first few days of April to support RCT-1 near al-Kut, then supporting Task Force Tarawa at al-Amarah and an-Numaniyah, farther south of Baghdad. The squadron flew surveillance missions to support the many convoys heading through the city and monitored large demonstrations there. On 25 April, it moved to ad-Diwaniyah.[56]

The nonstop flying during the final days of the assault into Baghdad meant the maintenance crews at Salman Pak worked around the clock to repair and rearm attack helicopters for the fight. Referred to as Yankee FARP, aviation ground support dispensed more than 127,300 gallons of fuel to 386 aircraft during air operations there, and KC-130s delivered pallets of supplies.[57] To show his appreciation for what MWSG-37 had done to expeditiously secure and prepare the airfield for operations, General Amos arranged a special delivery of fruit on one of the KC-130 flights.[58]

Urban Close Air Support over Baghdad

As the Baghdad offensive loomed, the likelihood of conducting close air support in an urban area became almost certain. Thus, the aviation asset of choice was the attack helicopter because of its maneuvering capabilities in confined spaces at low altitude and the proficiency of their aviators in engaging targets. The helicopters carried a near-precision weapons load of tube-launched, optically tracked, wire-guided missiles and 20mm guns.[59] Based on the rules of engagement, which required positive identification of enemy targets before engaging to minimize civilian casualties and damage to civilian property, helicopter gunships seemed the ideal platform. While air support may have been welcomed by ground forces, gunship use presented an extremely high risk for the aircraft and their crews who would be exposed to enemy small-arms and rocket fire from rooftops and alleyways. In addition, spotty communications with FACs and limited suppression of enemy fire increased the overall risks. Planners wanted to avoid a Mogadishu-type situation* with downed aircraft in downtown Baghdad, so reducing risks and improving the safety of aircrew were top priorities.

General Amos planned to use attack helicopters cautiously during the final offensive, accepting that they had suffered extensive wear-and-tear and battle damage in previous combat. Although willing to commit the assets, he equally understood and expressed that Marine aviation tactics required major modifications to counter the enemy's tactics in an urban area. As the general clearly stated, "We can't just willy-nilly take Cobras and Hueys at 150 or 200 feet over the top of the grunts [ground troops] while they advance deeper into the core of Baghdad, an urban setting with over five million people, just to hover with the ground forces. If we did, we would lose our pilots and the aircraft; and once you've lost them, they're irreplaceable. So we, 3d MAW, changed tactics."[60]

I MEF Air Officer, Colonel Paul K. Hopper, had been deeply involved in developing the urban air attack plan on Baghdad. Among the continuously changing battlefield dynamics to which he had to adapt was the modification of the size and shape of the restricted operation zone around Baghdad by the Coalition forces air component commander. The zone was a control measure similar to the air control battlefield coordination line that distinguished unit boundaries when coordinating combined arms and fire support plans. The boundary change benefited the Marine Corps because the Air Force air support operations center was unable to control the I MEF area of operations. As a result, the control agency turned over I MEF aircraft to the

*During the joint Operation Restore Hope in Mogadishu, Somalia, in 1993, Army helicopters were shot down in urban areas, leaving aircrews and soldiers surrounded by the enemy in a hostile city with little or no chance of escape or rescue.

direct air support center, which pleased the MACG-38 staff and controllers. During the Baghdad air campaign, I MEF eased air space coordination by offering all Marine Corps sorties in direct support of the joint force air component commander for the remainder of combat operations.[61]

General Amos requested that a team of instructors deploy from MAWTS-1 and be assigned in various positions to support the air campaign. Even the squadron's commanding officer, Colonel Raymond C. Fox, volunteered to deploy. One morning a few days before the Baghdad offensive, General Amos assigned Colonel Fox to develop revised tactics based on a better understanding of the enemy after an-Nasiriyah. His task was to offer a better method of employing Cobras and Hueys to support Marines fighting in an urban environment. Responding to the urgency of the moment, he devised a box formation in which two Cobras would lead with a few thousand feet of separation followed by two Hueys. Because Cobra crews could see the enemy in front and to their sides but not to their rear, the Hueys would cover their "six" with the extra eyes and weapons of each

Photo by SSgt Michael R. Picklo. Defense Imagery 030514-M-TI721-002

The 33d Commandant of the Marine Corps, Gen Michael W. Hagee, is briefed by Col Raymond C. Fox, commanding officer of MAWTS-1. Fox was one of four battle captains in the tactical command center during Operation Iraqi Freedom.

A ground aviation support crew from MWSS-373 loads Hellfire missiles while refueling an AH-1W Super Cobra from HMLA-276 in Tikrit.

Photo by LCpl Nicholous L. Radloff. Defense Imagery 030414-M-5654R-026

The EA-6B Prowler Disrupts Enemy Artillery Attack

Although CentCom determined that enemy electronic warfare was not a threat, Prowlers flew jamming and psychological operations missions to further confuse their command-and-control network. The Marine Corps deployed two of its four Marine Tactical Electronic Warfare Squadrons (VMAQs)—VMAQ-1 and VMAQ-2—from 2d MAW out of Cherry Point, North Carolina, to Prince Sultan Air Base, Saudi Arabia, in February 2003. During seven weeks supporting Operation Southern Watch and subsequently OIF, VMAQ-1 logged 1,129 combat hours and flew 197 combat sorties while VMAQ-2 flew more than 1,000 combat hours.[62]

On 7 April, RCT-1 never realized that a lone Prowler most likely prevented several unit causalities because of the aircraft's sophisticated capabilities. As RCT-1's lead unit prepared to cross the southern bridge over the Diyala River near the southeastern edge of downtown Baghdad, an operator from a nearby radio battalion intercepted a series of transmissions revealing enemy plans to destroy the bridge with artillery. The information was quickly forwarded to the regimental air officer, who sent an electronic warfare support request to the direct air support center. The air controller quickly located an EA-6B and assigned it to disrupt Iraqi communications, while also coordinating friendly troop positions. Although Prowler joint air requests typically entailed longer response times and additional levels of command-and-control protocols, the quick reactions by all involved—the communicator, air officer, air controller, and aircrew—most likely avoided casualties by thwarting enemy plans.

A strong advocate of electronic warfare capabilities, Lieutenant Colonel Wade C. Hall, a former VMAQ-3 commanding officer, served as the 3d MAW electronic warfare officer on the tactical air command center battle staff. Familiar with the circumstances, Colonel Hall believed this mission saved Marines despite their not knowing that the Prowler was patrolling and protecting them from overhead.[63]

of their two door gunners at the ready. General Amos called this formation "smart flying" because he wanted the crews to return safely and fly again the next day in fully operable aircraft.[64]

Despite aircraft strengths and pilot skills, Cobra gunships repeatedly received enemy fire when flying urban close air support missions. By the time I MEF reached Baghdad, almost 40 percent (or 22) of the deployed Cobras and 2 Hueys had sustained battle damage.[65] Commonly cited by ground Marines as their close air support weapon of choice, the Cobras' battle damage came primarily from small-arms fire that hit rotor blades, hydraulic systems, and fuel cells.[66]

As 1st MarDiv moved closer to Baghdad, General Natonski's Task Force Tarawa veered toward the eastern part of Iraq to confront the *10th Armored Division* in the vicinity of al-Amarah. General McKiernan viewed the Iraqi force as a potential threat that had not been encountered during combat operations. On 8 April, Task Force Tarawa coordinated Coalition air strikes on the remaining elements of the division.[67]

A Shattered Statue and a Regime Change

On 9 April, Marines arrived in Baghdad after a rapid move from northern Kuwait that took three weeks. Senior commanders, fully expecting a seven-week trek to enter the capital, were pleased. Planners expected a fierce fight and assumed the city would be heavily defended because it represented Saddam's economic, political, and military power base. With the conventional enemy threat severely diminished along the 400-mile route from Kuwait, Marine air-ground task force efforts focused on quickly seizing the city. Ridding Baghdad of Saddam loyalists was one of I MEF's objectives.

The three regimental combat teams and the 11th Marines divided the eastern part of the city into four discrete battle sectors. The regimental commanders each exercised central authority to secure their respective areas and protect the civilian population. The intent was to neutralize remnants of enemy resistance with continual foot and vehicle patrols and begin meeting with local governmental officials and civilians who could

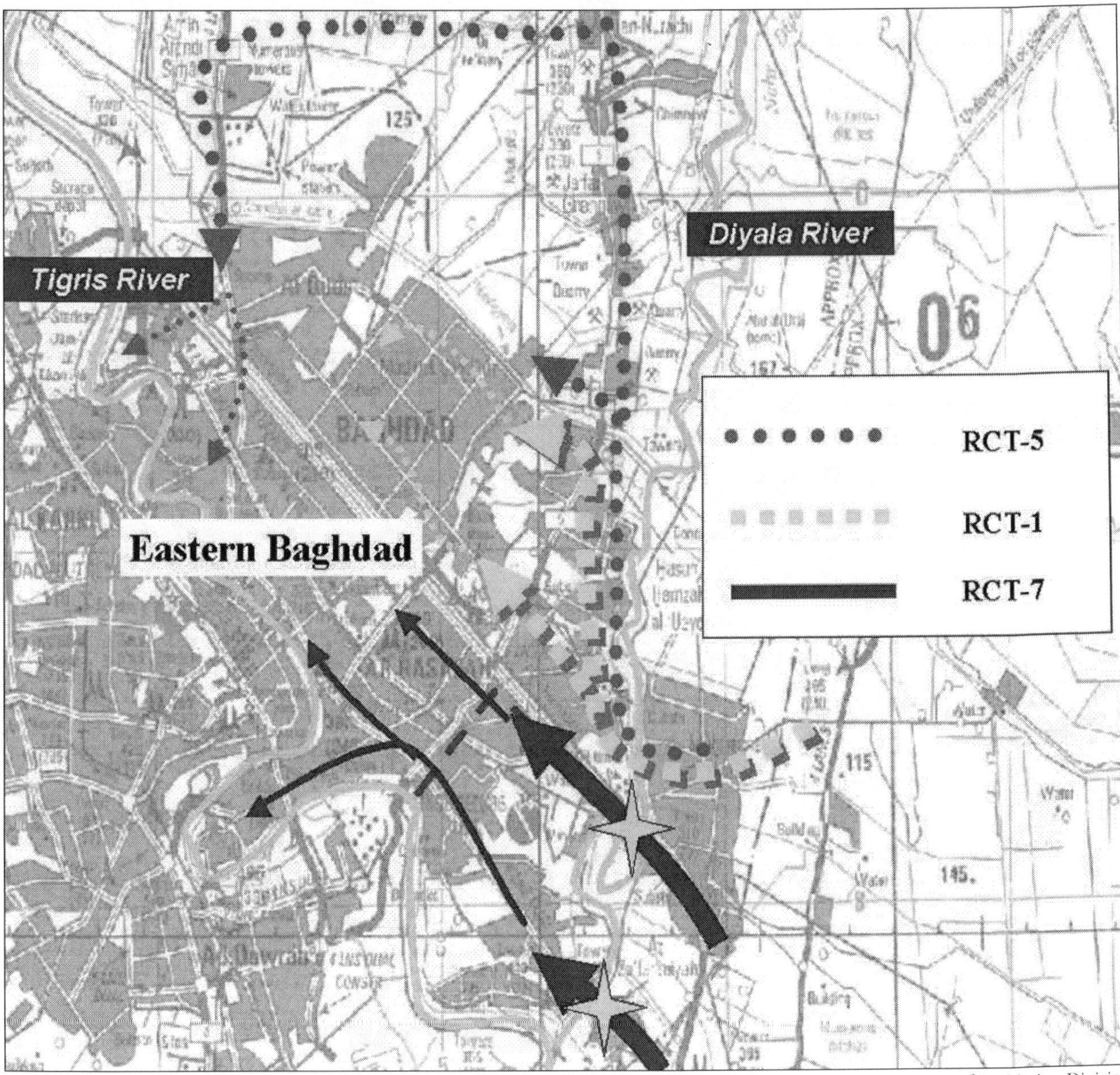

Map courtesy of 1st Marine Division

identify Saddam loyalists.[68] RCT-5 encountered the most activity with lingering resistance from fedayeen and paramilitary fighters as it approached Baghdad from the north. Concurrently, the Army controlled the area west of the Tigris River. Together, the Marines and Army reduced lingering enemy resistance by projecting a visible presence on the streets of Baghdad. General Mattis allowed subordinate commanders to determine whether their units would attack the city while wearing bulky protective suits. The sense was that Saddam would only use weapons of mass destruction as Coalition forces neared the city, not where he and the core of his elite resided.[69]

Crowds of civilians expressed welcome by waving and giving thumbs-up gestures. Cheers and chants of praise welcomed the Marines who first drove tactical vehicles into the core of Baghdad. When an RCT-5 advance patrol moved cautiously toward a prison compound, they were pleasantly surprised by their discovery. Instead of enemy resisters, a large gate opened and more than 100 children clad in threadbare clothing fled from the building toward the Marines. Several children ran with crossed arms as if they may have been accustomed to handcuffs or some type of restraints; however, they quickly changed their body language and joined nearby Iraqis with echoes of "thanks" and "good."[70]

Later that day, Marines witnessed history in the

Photo courtesy of 1st Marine Division

Cpl Edward Chin of Company B, 1st Tank Battalion, drapes an American flag over the head of a statue of Saddam Hussein in Baghdad's Firdos Square. Moments later, he was ordered to replace it with an Iraqi flag.

making as Iraqi civilians began taking ownership of their country that had been ruled for almost 24 years by a brutal dictator. Iraqis gathered in the streets to embrace their newfound freedom by toppling a large statue of Saddam Hussein overlooking Firdos Square in downtown Baghdad. Iraqis used shoes, sledgehammers, cardboard boxes, sticks, and even garbage to topple the statue that depicted the Iraqi president standing tall in a civilian suit, right arm raised in a wave to his people. They flailed at it for a while before a Marine M88 tank retriever tore it down. The Iraqis then attacked the fallen statue.

On 10 April, RCT-5 seized Azimiyah Palace, one of Saddam's opulent dwellings. The 1st Battalion, 5th Marines, also received two fragmentary orders to search a site suspected of holding captive prisoners of war and seize the Imam Abu Hanifah Mosque where Saddam reportedly had been sighted only hours earlier. RCT-5 took fire from small arms and RPGs. One Marine was killed and 42 others were injured, but by evening Marines had killed hundreds of fedayeen fighters.[71] The 3d MAW's close air support and medical evacuations helped reduce initial resistance in this part of the city.[72] A major milestone was achieved by capturing the capital without a massive counterattack from conventional Iraqi military forces. Marines cautiously patrolled the streets of Baghdad as fixed-wing aircraft and tactical helicopters patrolled unchallenged above.

On 11 April, 3d MAW flew only 174 missions

Photo courtesy of Col Charles J. Quilter II

SgtMaj John L. Estrada of 3d MAW sits in a chair in one of Saddam Hussein's opulent palaces.

with just 33 percent in direct support of Air Force General Moseley, the Coalition forces air component commander. Although a slowdown was nearing, the air wing kept its airborne FAC patrol on standby for immediate Marine close air support missions. After 23 days of sustained air operations, Baghdad collapsed quickly. Once the city was secured, Marine aircrews reverted to supporting I MEF, but with a noticeably scaled-down schedule. The Moseley-Amos prewar handshake had surpassed expectations and defined an improved strategy for managing joint airspace. Although conventional military firefights subsided as Marines established a visible presence, new threats emerged inside the capital that were unexpected in scope, presenting a host of new challenges. The exodus of Saddam Hussein's government and civic leaders provided a ripe environment for civil unrest fueled by street gangs, looters, and irregular paramilitary resistance. Baghdad quickly became a city of transition and turmoil.

Part IV
Transition to Security and Stability Operations

Chapter 11
Regime Removal

Regime Removal

Seizing Baghdad had always been a prerequisite for implementing regime change, but civil unrest and looting had to be controlled, public utilities had to be restored, and security had to be in place before Coalition forces could offer humanitarian aid to the Iraqi population. Iraqi government workers' failure to report to work further compounded this dynamic situation. It appeared that most of the once-loyal supporters of Saddam Hussein's powerful regime seemed to have simply vanished, while looters, the unemployed, and opportunists took their places, stealing and stockpiling food and supplies from vendors and unprotected storefronts.[1]

General Conway, the I MEF commander, was reminded of the 24-hour shift in operations after a civilian suicide bomber blew himself up, injuring five Marines from RCT-5 in downtown Baghdad. Only blocks away, Marines also faced random firefights from paramilitary fighters while carrying out security activities, such as protecting international Red Cross facilities. It was a picture book scenario of the "three-block war" concept, introduced by the 31st Commandant of the Marine Corps, General Charles C. Krulak. Describing a post–Cold War battlefield, the Commandant envisioned Marines performing a variety of simultaneous missions ranging from hostilities to peacekeeping and humanitarian relief.[2] General Conway provided a first-hand view of Baghdad just one day after Coalition forces

A Marine M1A1 main battle tank with Company C, 1st Tank Battalion, drives through downtown Baghdad en route to a new assembly area just five days after the fall of the city.

Photo by Sgt Paul L. Anstine II. Defense Imagery 030414-M-UR305-021

Photo by LCpl Christopher H. Fitzgerald. Defense Imagery 030420-M-YA501-002

Flying over Baghdad on 20 April 2003, a CH-46E Sea Knight from HMM-165 flies past the Iran-Iraq War Memorial without encountering any antiaircraft or small-arms fire.

A Marine Corps explosive ordnance disposal unit destroys an enemy ammunition cache at a Republican Guard *base in Iraq during Operation Iraqi Freedom.*

Photo by Sgt Mauricio Campino. Defense Imagery 030417-M-DA916-024

Photo by Maj Kathleen A. Hoard. Defense Imagery 030424-M-RL900-010

Marine Corps LtCol Dave Long (center rear), Army SFC T. Bucci, and Army Maj Phil Osterli (right)—with members of I MEF and 3d Civil Affairs Group—give soccer balls and cookies to the head instructors at a newly renovated high school in Umm Qasr.

seized the city: it was "a sobering reminder . . . that much work remains to be done." The Marines "must . . . quickly transition to Phase IV Stage B civil-military/humanitarian assistance operations to stabilize the situation and demonstrate our resolve to . . . the newly liberated . . . [Iraqis]."[3]

Although various security and stabilization options had been debated for months at CentCom headquarters in Tampa, the final approach for conducting a regime change and post-hostility operations was not as deliberate or as clearly defined as the other three operational phases—preparation, battle shaping, and decisive action—of OPLAN 1003V. This revelation was mostly attributed to the fact that the plan did not address details about civil-military operations that would require integrated military, economic, and political solutions and collaboration among U.S. Department of Defense, other government sectors, and Coalition partners. Instead, it centered on seizing the southern Iraqi oil fields, removing the regime, and locating weapons of mass destruction. In fact, the I MEF sequel to OPLAN 1003V that addressed post-hostility operations was dated 20 April—almost two weeks after the fall Baghdad. That document described I MEF's area of operations as bounded by the Iran-Iraq border on the east and north and by the al-Faw Peninsula in the southern part of the country; I MEF now covered nine governates within southern Iraq.[4]

Security and stabilization operations in Baghdad proved to be one of the most difficult phases of the campaign to implement, particularly restoring public utilities such as electricity, water, and telephone service, along with providing humanitarian assistance for the Iraqi people. Assessing the cha-

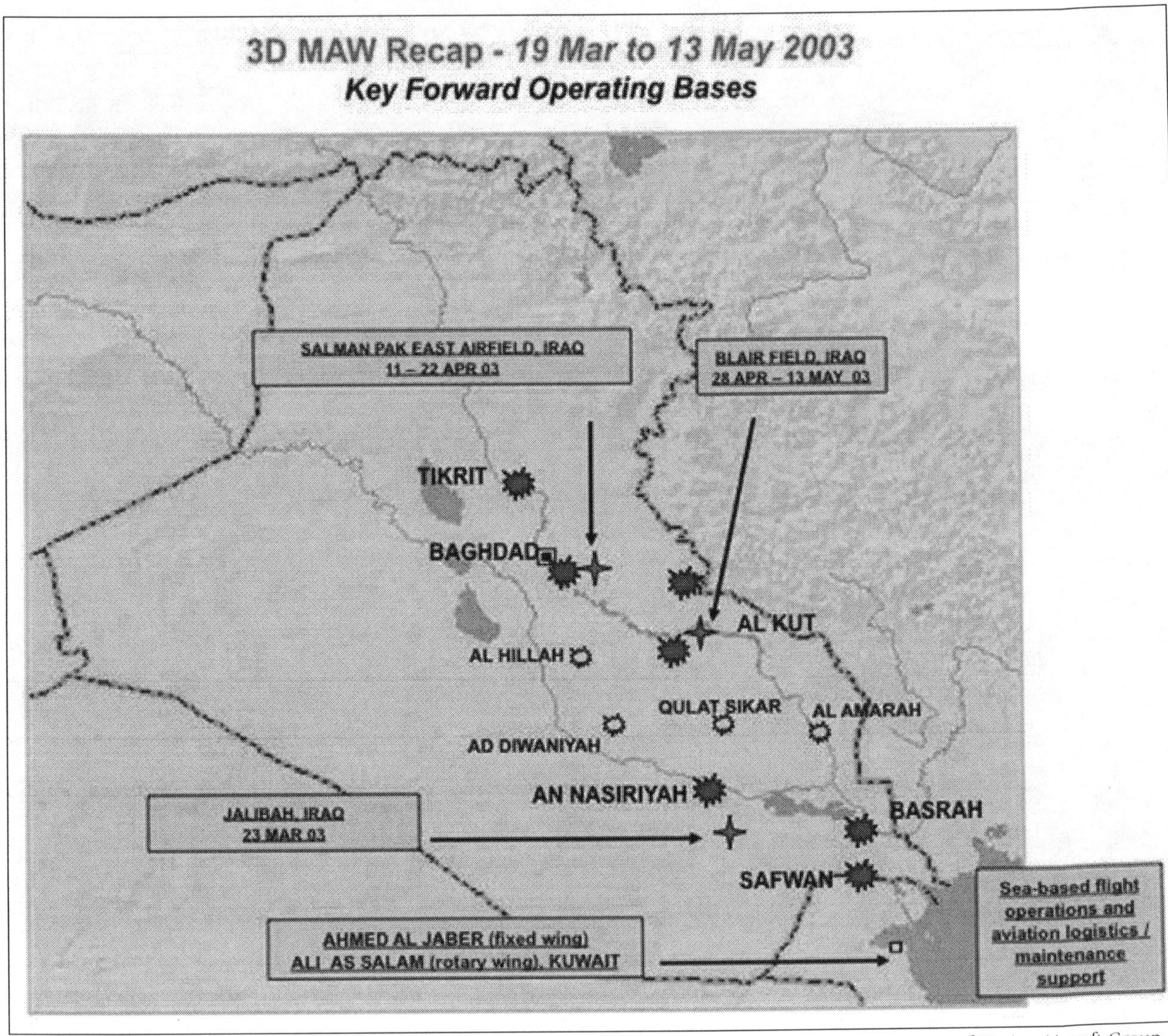

Map courtesy of Marine Aircraft Group 29

os, General Mattis, 1st MarDiv commander, chose not to wait for higher headquarters direction and established a civil-military operations center in downtown Baghdad at the Palestine Hotel.* With the 11th Marines providing oversight and with assistance from RCT-1 and RCT-7, the center held daily meetings with former and current Iraqi civic leaders and nongovernmental organizations, including the International Federation of Red Cross and Red Crescent Societies (IFRC)** and Doctors Without Borders.[5] Small strides were achieved in meeting basic survival needs—food, shelter, and clothing—in a city with six million residents. The operations center became a magnet of action and hope, attracting Iraqis wishing to practice their newly discovered right of free speech. In essence, it was a small band-aid for the wounds of war until the Army's 3d Infantry Division conducted its turnover on 21 April.[6]

Unsure of the future operational tempo, 3d MAW continued a 24-hour presence over Baghdad with on-call close air support. This continued until 14 April, when combat activities in Baghdad slowed significantly. Fortunately, strict adherence to rules of engagement by the wing during combat operations paid gigantic dividends as the narrow scope of collateral damage reduced the need for extensive reconstruction efforts.

*Typically, the Marine Corps does not engage in such large-scale civil-military activities, but Gen Mattis understood that the Iraqi people had been oppressed for years and needed an immediate sense of normalcy in their lives.

**The IFRC is most commonly known as the International Red Cross. IFRC was founded in Paris in 1919 in the aftermath of World War I, and its early efforts focused primarily on prisoners of war and other combatants. Today, the IFRC is the world's largest humanitarian organization, and it focuses on vulnerable populations across the globe.

Expanding Operations North to Tikrit

As 1st MarDiv gained control of Baghdad, General Conway issued a formal warning order on 11 April stating that elements of I MEF would proceed farther north to Kirkuk. The continuing offensive beyond Baghdad was to seek suspected enemy remnants and capture the northern oil fields. Sustainment operations would be stretched even further with the addition of another 150 miles to the strained logistical flow. Regardless, Generals Conway and Mattis relied on General Amos's 3d MAW to maneuver ahead of and with the ground forces, once again. The wing's fleet of KC-130s along with its CH-53 Super Stallion and CH-46 Sea Knight cargo and transport helicopters and MWSG-37 once again ramped up their operational tempo.

The rapid successes the next day when Kurdish and U.S. Army Special Forces seized the northern oil fields in Kirkuk altered those plans. Although I MEF was still ordered to deploy north of Baghdad, it was now directed toward Saddam's hometown of Tikrit. It had been a stronghold of support for him and was likely harboring loyalists of the *Special Republican Guard* and paramilitary irregulars who would take a final stand against the Coalition forces.

Despite no extant plans to support air operations 100 miles north of Baghdad, General Amos's planners expanded 3d MAW's area of operations to Tikrit. This meant gaining control of aircraft in a new airspace while redeploying aircraft, equipment, and wing personnel. The I MEF air branch, headed by Colonel Hopper, orchestrated the bulk of this effort, focusing on a way for military and civilian aircraft to share the airspace. The redeployment of air operations near Tikrit set a priority for vital close air support and medical evacuations while developing a management plan to accommodate commercial aviation use for future reconstruction efforts.[7]

Third MAW attributed this accomplishment to MWSG-37, particularly its engineer support Marines. At nearby an-Numaniyah airfield, aviation ground support crews made essential runway repairs, repeating similar tasks they had performed on multiple occasions throughout 1st MarDiv's assault in Iraq. The teamwork paid off within 48 hours when KC-130s began landing safely, initiating a renewed flow of supply support. One of the first flights landed with General Conway on board and the aircrew delivered critical light armored vehicle repair parts.

On 11 April, MAG-13's commanding officer, Colonel Savarese, led one of the first Harrier sections north of Baghdad into the newly built

A Marine Corps KC-130's cargo, including MREs, is unloaded for ground troops at an-Numaniyah Airfield.

Photo courtesy of Col Charles J. Quilter II

an-Numaniyah airfield, code-named Three Rivers FOB.[8] Deep inside Iraq, Colonel Savarese assessed the airfield's condition. If the condition was acceptable for flight operations, the group would have much more time on-station and could facilitate more responsive strike coordination and armed reconnaissance missions. Over the rest of the month, MAG-13 squadrons logged 327.5 hours and more than 165 sorties at Three Rivers supporting the ground combat element.[9]

Organizing Task Force Tripoli

In response to the newly assigned offensive, General Mattis formed a joint task force for the Tikrit mission—Task Force Tripoli—led by assistant division commander and newly promoted Brigadier General John F. Kelly. It consisted of three light armored reconnaissance battalions reinforced by artillery from 11th Marines, aviation from 3d MAW, ground logistical support elements, and a Navy SEAL team. Named in honor of Lieutenant Presley O'Bannon's 1805 desert march, Task Force Tripoli headed north with less than 12 hours notice with a force of 4,000 troops and about 600 wheeled vehicles.[10] General Kelly's orders were to rid the town of any remnants of Saddam's leadership. Given the urgency to continue the momentum north, there was no time for excessive planning or a detailed operational order, but the force was well equipped with a varied assortment of combined arms—particularly 3d MAW and its precision munitions.

On 12 April, a mix of KC-130s, CH-46s, and CH-53s supported the task force while Cobras and Harriers provided armed escort and close air support overhead and 3d FSSG trailed close behind.[11] MAG-29 supplied the bulk of helicopter assets for the push up Highway 1 to Tikrit because its composite mix of tactical, transport, and utility helicopters offered "one-stop shopping." Still operating until late April from Salman Pak, MAG-29 detached from 3d MAW and returned to Task Force Tarawa as its aviation combat element, flying primarily from Blair Field.

Navy corpsman Derek Fingland, assigned to MALS-29, stands watch next to a CH-46E Sea Knight from HMM-162 as it refuels at an-Numaniyah.

Photo by MSgt Howard J. Farrell. Defense Imagery 030406-M-KS959-135

Elements of MACG-38 joined MWSG-37 by ground convoy—encountering small-arms fire along the 100-mile route—to set up tactical communications for the assault at a FARP dubbed Tikrit South. The FARP was operational until 21 April, when 3d MAW's operational tempo slowed and air operations began moving south. At one point, the convoy was flagged down by attack helicopters that desperately needed fuel. Without hesitation, air controllers and ground aviation crews formed a "hasty" FARP for the Cobras near Samarah, about 50 miles south of Tikrit, further demonstrating the ability of the air-ground team to stretch its logistical support beyond expectations.[12] As Task Force Tripoli neared Tikrit's city limits, they initially passed through friendly sectors with cheering and waving Iraqis, but sentiments became more hostile closer to the city core. An abundance of Saddam Hussein statues, murals, and paintings clearly marked the town of his birth.[13]

Rescue Mission of Seven American Soldiers

On 13 April, Marines made headlines when they joined efforts to rescue seven Americans from Samarah, a small village about 75 miles north of Baghdad. Members of Task Force Tarawa from 3d LAR Battalion were advancing toward Tikrit, intending to keep civilian traffic from interfering with tanks also headed to the city, when they received intelligence about American prisoners of war located nearby. With little information, an Iraqi informant escorted the Marines to a building where they found, under guard, U.S. Army soldiers believed to be members of the 507th Maintenance Company who had been ambushed during the battle of an-Nasiriyah a few weeks earlier.

The Marines rescued the soldiers without a fight from the guards and moved them to a waiting CH-46 from HMM-165, which airlifted all seven injured prisoners—five from the 507th and two Army attack helicopter pilots—to safety, landing at an-Numaniyah airfield south of Baghdad. Farther south, the KC-130 aircrew from VMGR-452, a reserve squadron from Stewart, New York, had just finished loading pallets of mail for delivery to forward sites in Iraq when they received different orders—to rescue American soldiers. The aircraft commander, Major William H. Holmes, had the mail cargo quickly unloaded and the cabin outfitted to accommodate wounded passengers.

At an-Numaniyah, the former captives were transferred to the KC-130 and flown directly to Kuwait International Airport. As the aircraft taxied to its parking spot, the aircrew hung an American flag from the back of the plane.[14] The repatri-

A light-enhanced photograph shows an explosion resulting from Coalition aircraft attacking hostile forces in Saddam Hussein's hometown of Tikrit on 14 April 2003.

Photo by Sgt Nicholas S. Hizer. Defense Imagery 030414-M-UE267-009

Photo courtesy of Cpl Michael Leitenberger

Marine Corps Cpl Christopher Castro (seated left) and LCpl Curney Russell II (standing right) speak with U.S. Army soldiers SPC Joseph Hudson, CWO-2 David S. Williams, and CWO-2 Ronald D. Young Jr. about their rescue on 13 April 2003. This photograph was taken on the KC-130 commanded by Maj William H. Holmes while en route from Iraq to Kuwait City.

ated Americans cheered as they exited the aircraft alongside their Marine liberators and were greeted by a host of reporters. Corporal Richard S. Austin, the KC-130's loadmaster from Kansas, recalled, "Being a part of the POWs' return was a definite highlight of serving in Operation Iraqi Freedom. That was the best mission we've had since this operation began. It feels great to be a part of their return to freedom. It's an important and memorable highlight for the crew. We will never forget this mission."[15]

The mission was an inspiring morale booster at 3d MAW, especially for those Marines who were closely involved. It demonstrated the lengths to which U.S. forces will go for their fellow Americans. In many ways, the mission became one of the defining events that marked the end of the combat air campaign

Post-Hostilities, Diplomacy, and Humanitarian Assistance

By mid April, the U.S.-led Coalition had successfully removed Saddam's dominant Baath Party from power in Baghdad and Tikrit, ending decades of brutality against the populace and defiance of UN Security Council resolutions. Although Saddam was not immediately captured after Baghdad fell, the now-leaderless nation fell under Coalition control.* On 21 April, a transitional government—the Coalition Provisional Authority—was established under the leadership of retired Army Lieutenant General Jay M. Garner.** General Garner's tenure, however, was short-lived. On 6 May 2003, he was replaced by 23-year State Department veteran L. Paul Bremer III, a career diplomat who was proclaimed one of the world's leading experts on crisis management, terrorism, and homeland security. Unfortunately, Bremer made two decisions during his tenure that were questioned by military leaders and the media at the beginning of phase

*Until elections were held in 2004, Iraq's interim government relied on the Coalition Provisional Authority to provide assistance and advice regarding administrative, political, economic, welfare, and reconstitution matters.

**The UN Security Council also encouraged other countries to assist by authorizing more than 60 nations to compete for contracts to rebuild Iraq.

Photo by LCpl Christopher G. Graham.
Defense Imagery 030604-M-IW639-022

A group of Iraqi youths show their support at the al-Moter Secondary School in the town of al-Hillah, where local Iraqi officials along with U.S. Marines and soldiers assigned to civil affairs units paid wages for teachers to support school activities and boost morale.

IV operations: removing members of the Baath Party from public service—de-Baathification—and disbanding the Iraqi military. Bremer's rationale for these decisions was summarized in a personal account published by the *Washington Post* four years later. His article compared Saddam Hussein's regime to that of Nazi Germany. In Bremer's view, Iraq also had no option other than to abolish the Baath Party and the Iraqi Army if any hope would be offered for the Iraqi people.[16]

After securing Baghdad and ridding Tikrit of Saddam's loyalists, General Conway recalled a conversation with senior Army commander General McKiernan about what the Marines would do next. Feeling confident about the Marines' accomplishments, General Conway noted, "'Marines don't do nation-building.' My boss quickly told me to get my Marines south and conduct reconstruction efforts until I can get YOU relieved."[17] In the end, a small Marine presence remained in Iraq for nearly five more months, conducting security and stability activities.

Meanwhile, as the 1st MarDiv was busy securing Baghdad and Tikrit, Task Force Tarawa maintained security in southern Iraq, ranging from an-Nasiriyah along Highways 1 and 7, and orchestrated humanitarian relief. Far to the north on the Iraq-Turkey border, Major General Henry P. Osman, II MEF's

Cpl Richard James, assigned to MWSS-373, offers a bottle of water to an Iraqi boy at a FARP in northern Iraq.
Photo by LCpl Jason L. Andrade. Defense Imagery 030430-M-5972A-017

commander, had deployed during the first few days of combat operations without much notice. The expeditionary force was anticipating large-scale humanitarian efforts that would be needed after Baghdad fell and phase IV operations began.

With mentorship from retired General Anthony C. Zinni, General Osman formed a small headquarters and established a military coordination and liaison command with the sole purpose of reinforcing political stability on the northern border. He later recalled,

First, it started out as this . . . joint task force to execute a kind of Provide Comfort type of mission. But as the thing began to develop, I realized I was going to be in a liaison coordination role with whomever would show up, up there, coordinating efforts between the militia that the Kurds and the Peshmerga [Kurds who fight for a free Kurdish state] had, about 75,000 between them. And then coordination between the various factions was really critical, and that I think became . . . our main effort; however, in the end, a massive . . . humanitarian effort was really needed.[18]

Chapter 12

Redeploying the Air Wing

Closing Kill Boxes and Winding Down

A few days after the fall of Baghdad, 3d MAW closed its kill boxes, which signified a drastic slowdown in combat-related air support requests. By 15 April, 3d MAW was flying 45 percent fewer missions than at the peak of combat operations because enemy targets either had been destroyed or no new targets were identified. On 18 April, Marine air operations waned even more when the Army's 3d Infantry Division, 358th Civil Affairs Brigade, deliberate assessment team relieved 1st MarDiv of eastern Baghdad responsibilities, and the 4th Infantry Division relieved Task Force Tripoli in Tikrit. These two changes in areas of responsibilities enabled 1st MarDiv to head south toward its new command post in ad-Diwaniyah.[1] Three days later, the airborne direct air support center halted 24-hour operations, ending with 850 flight hours while directing 2,261 helicopter and 210 fixed-wing missions.[2]

On 1 May 2003, President Bush, outfitted in full flight gear, landed on the fight deck of the USS *Abraham Lincoln* (CVN 72) and stated to the nation that major combat operations in Iraq were over. This announcement initiated a flurry of activity among the Marine air-ground team that focused on planning for an arduous, yet welcome, redeployment home to the East and West Coasts. The manpower division of Headquarters Marine Corps issued Marine Administrative Message 228/03, which ended its stop-loss action, thus

President George W. Bush poses for a photograph with sailors on board the aircraft carrier USS Abraham Lincoln *(CVN 72). While on board, the president addressed and praised the sailors who were returning from a record-breaking 10-month deployment in support of Operation Iraqi Freedom.*

Photo by PO3 Tyler J. Clements, USN. Defense Imagery 030501-N-LN737-087

Photo by LCpl Andrew Williams. Defense Imagery 030913-M-UW798-005

In September 2003, LtCol Thomas F. Dietrich (right), commanding officer of VMGR-234, shakes the hand of one of his Marine reservists after returning to home base in Fort Worth, Texas, following a long deployment in support of Operation Iraqi Freedom.

further confirming that personnel were no longer restricted from requesting separation, retirement, or transfer actions, and administrators could continue with normal succession planning.

Under Deployment Order 164—the same one that sent 3d MAW to Kuwait and Iraq five months earlier—General Amos's staff devised a phased redeployment schedule that began on 5 May with movement of squadrons and aircraft to Marine Corps Air Stations at Miramar, Yuma, Cherry Point, Beaufort, New River, and numerous reserve sites. On the opposite end of the spectrum, planners coordinated retention of a much smaller force in Iraq, consolidating FARPs to bases at al-Hillah, al-Kut, and Tallil. Ahmed al-Jaber Air Base was essentially shut down, and 3d MAW moved its command element to Ali al-Salem.

MAG-16 began shuttling from Jalibah and Ali al-Salem to the USS *Boxer* (LHD 4) on 9 May, transferring HMM-161 to its parent command, the 15th MEU (SOC); HMM-165 remained with 3d MAW (Forward). Fixed- and rotary-wing squadrons returned to ATF-West and ATF-East with liberty stops in Australia and Portugal, respectively, as remaining 3d MAW personnel boarded military and commercial flights for their journey home.[3]

Before units departed, the 3d MAW band played in a series of performances during the final days of May, scheduling several events at Camp Commando and Ali al-Salem in Kuwait and a more somber Memorial Day ceremony at Ahmed al-Jaber Air Base on 26 May. During combat operations, the band supported perimeter security tasks and assisted with duties at Ahmed al-Jaber.

On 6 June, General Amos left Iraq, entrusting assistant wing commander General Robling with the final phase of Iraqi operations. Assuming duties as 3d MAW (Forward) commanding general, Robling retained the bulk of MAG-39 to continue supporting I MEF with security and stability activities. It was augmented by a detachment of KC-130s, a Harrier squadron from VMA-214, and MWSS-371 for aviation ground support. By 30 June, 3d MAW had decreased its personnel

Photo by Col Charles J. Quilter II

The 3d MAW band performs prior to redeployment to home base.

Marines from VMGR-234 conduct a preflight inspection of a KC-130 Hercules at Baghdad International Airport. The desert sand and dust were particularly hard on engines and propellers, yet maintenance crews kept the aircraft flying.

Photo by LCpl Andrew Williams. Defense Imagery 030807-M-UW798-009

strength within the CentCom area of operations to 3,500 Marines and sailors; however, the urgency and intensity of varied tactical, logistical, and goodwill missions were nonetheless just as critical during the remaining months of the operation.[4]

The smaller Marine air-ground task force remained in place and conducted a range of missions. By 15 June, 3d MAW distributed its fleet of 106 aircraft across five sites: MAG-39 main air operations were based at Ali al-Salem; KC-130s and Harriers were at Ahmed al-Jaber; and smaller detachments of helicopters colocated with Army medical evacuation helicopters at Tallil, ad-Diwaniyah, and al-Kut.[5] Daily sorties included medical evacuations, logistical resupply, dignitary transportation, armed reconnaissance patrols, and on-call or strip alerts. The fleet of Super Stallions that remained behind was instrumental in moving cargo along resupply routes and in escorting a surge of dignitaries for tours of Iraq.

The hot summer months in Iraq were brutal for the air-ground team, yet Marines and sailors continued to dispense much-needed medical supplies and services along with deliveries of educational items to remote areas across Iraq. All of these efforts

Photo by LCpl Andrew Williams. Defense Imagery 030903-M-UW798-006

On the same day that the Polish-led multinational division assumed control of the I MEF area of operations, 3d MAW held a promotion ceremony at Camp Babylon, Iraq, for Sgt Brandon Whisler (center). Flanking Sgt Whisler are MajGen James F. Amos (left) and BGen Terry G. Robling (right).

Disposition of aircraft in Iraq as of 2 July 2003.

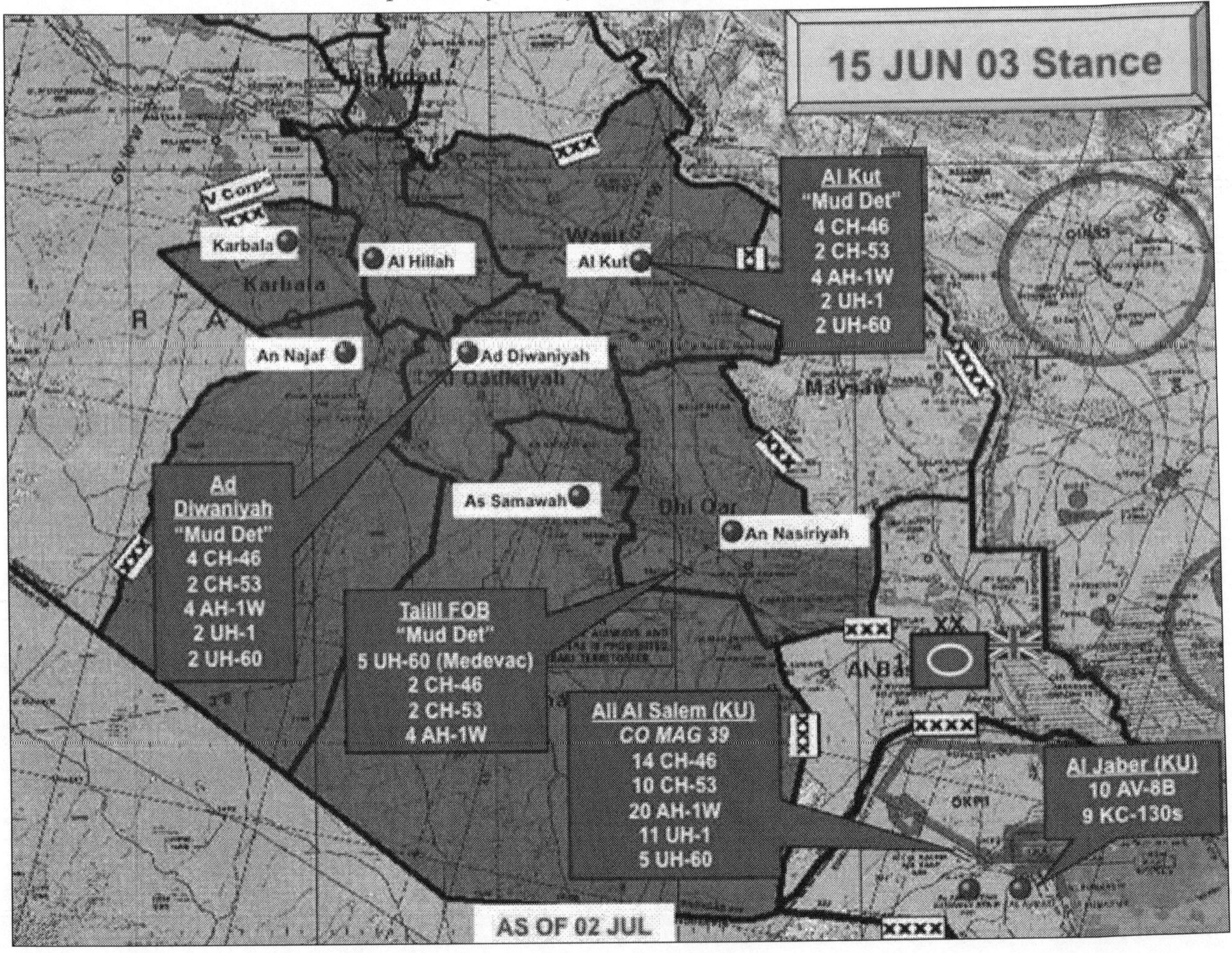

Photo by LCpl Andrew Williams. Defense Imagery 030819-M-UW798-016

BGen Terry G. Robling (third from left), acting commanding general of 3d MAW Forward, gives Gen William L. Nyland (second from left), Assistant Commandant of the Marine Corps, and dignitaries a tour of the facilities at Ahmed al-Jaber Air Base, Kuwait.

were aimed at promoting a sense of normalcy for the Iraqi people. After several planning sessions that had begun in May, the Polish-led multinational division officially relieved I MEF of its duties in its area of operations on 3 September 2003.[6]

Metrics

During combat operations, the U.S. Air Force reported that Coalition aircraft flew some 41,404 sorties, with Marine Corps aviation accounting for 23 percent.[7] Marine pilots and aircrews flew more than 9,800 missions, logging more than 35,000 flight hours in March and April alone.[8] The 3d MAW's metrics reflect its collective efforts despite adverse weather conditions and logistical hurdles. From January to April 2003, the wing continuously increased monthly flight hours, flying 361 sorties at its peak on 28 March.

During the advance to Baghdad, two 3d MAW aircraft mishaps marred an otherwise good safety record. On 30 March, a Huey from HMLA-169 crashed, killing three Marines and injuring one aircrew member shortly after takeoff.[9] Two aviators from HMLA-267 were also killed on 5 April when their Cobra gunship crashed about 30 miles southeast of Baghdad—it had flown into a tower during a combat mission.[10] By the end of hostilities, the wing reported only six class "A" aircraft accidents—five helicopters and one Pioneer UAV.[11]

Epilogue

After Saddam Hussein's regime was toppled and his hometown of Tikrit fell under Coalition control, the air war slowed considerably, allowing General Amos to reflect on the aviation combat element's 23-day journey to Baghdad. Marked with a trail of aviation accomplishments, 3d MAW deployed more than 15,000 Marines, positioned 435 tactical aircraft at land bases and at sea, offloaded support equipment and inventories with the aid of heavy-lift CH-53E Super Stallions in just 45 days, and extended air wing operations over 400 miles from permanent air bases in Kuwait and naval ships in the Persian Gulf. These feats showcased the wing as a supporting maneuver element for the ground combat element and as a separate maneuver element, frequently surmounting challenges and exceeding command expectations. As I MEF pushed toward Baghdad, the wing's Harriers and Hornets patrolled the eastern flank and kept sizeable Iraqi ground units contained and out of the main fight, while fixed-wing aircraft and helicopters provided close air support for the main effort, often in urban environments.

Marine aviation supported Marine infantry units, first and foremost, but also made sorties available to the Coalition air component commander and directly contributed to a larger scale Air Force–controlled air war. Indeed, General Amos recognized and appreciated the Marines' contributions and also credited a large part of 3d MAW's success to his Air Force advocate. He declared: "For the very first time in the history of Marine aviation, we got to fight the way we always wanted to fight. I will be forever grateful, as well as the Marine Corps, for Lieutenant General T. Michael Moseley's approach of fighting this war from an air commander's perspective."[1]

Directly supporting 3d MAW combat operations, the wing intelligence collections section processed more than 15,000 images and imagery-related products from reconnaissance aircraft, including RQ-2B Pioneers and EA-6B Prowlers. Marine AH-1W Cobra attack and UH-1N Huey gunship helicopters provided close air support and became particularly adept at engaging targets in confined areas, frequently using TOW missiles and 20mm guns while minimizing damage to adjacent areas. Additionally, CH-46 Sea Knight transport helicopters repeatedly provided casualty evacuation under fire during the height of hostilities, while CH-53s moved equipment, cargo, and troops throughout Iraq. The Corps' workhorse aircraft, the KC-130 Hercules, proved its dual capabilities once again, conducting aerial refueling missions—servicing between 65 and 70 aircraft and delivering an average of 400,000 pounds of fuel each day—and providing a heavy-lift capability.[2]

Without the mobility of MACG-38 air control agencies providing a safe airspace and the MWSG-37 providing a network of forward air operations, the wing could not have been as responsive supporting Marines on the ground. Employing multiple direct air support centers enabled the rapid direction of strike aircraft, both Marine and Coalition, to deep targets and for close air support in conjunction with rapidly moving ground forces. Without air-ground support staged at nearby FARPs and FOBs, Marine aircraft could not have been as responsive to targets of opportunity in support of expeditionary ground operations.

The aviation combat element forced Iraqi forces to make one of two choices: remain in position and fight, or flee and become a target of opportunity. In so doing, 3d MAW isolated and attacked enemy activities with precision-guided munitions within strict rules of engagement that minimized civilian casualties and infrastructure damage. Despite extreme logistical challenges, extended communications, brutal flying conditions, an aging fleet, and remote air operations, the aviation combat element helped defeat eight Iraqi divisions: the *11th*, *14th*, and *18th Infantry Divisions*; the *51st Mechanized Division*; the *6th* and *10th Armored Divisions*; the *Baghdad Republican Guard Infantry Division*; and the *al-Nidu Republican Guard Armored Division*.[3] The overall impact eventually destroyed the enemy's will to fight.

For the most part, 3d MAW Marines traveled home in the same mode of transportation that had carried them to Kuwait almost six months earlier. A few days before hostilities began, General Amos had shared his prewar thoughts in an e-mail with his Marines, "We will win this war and the respect of the Iraqi people . . . and we will do it honorably."[4] As predicted, 3d MAW Marines followed the commander's intent. They complied with strict rules of engagement and understood that the fight was against a brutal regime, not the Iraqi people.

On 3 November, 3d MAW stood in formation as the Presidential Unit Citation was presented to all those who served under the I MEF command. A mere four months later, General Amos led 3d MAW back to Iraq to support OIF II,* once again fully engaged and contributing to the Global War on Terrorism.

*OIF II began to counter the Iraqi insurgency with even stricter rules of engagement. It lasted until the official withdrawal of U.S. forces at the end of 2011.

Notes

Prologue

1. George W. Bush, "Remarks at Barksdale Air Force Base, Louisiana, on the Terrorist Attacks," 11 September 2001. Online by Gerhard Peters and John T. Woolley, *American Presidency Project*, http://www.presidency.ucsb.edu/ws/?pid=58056, hereafter Bush, "Remarks at Barksdale."
2. Conversations between Col Susan G. Sweatt and author, October 2001.
3. Alfred Goldberg, *Pentagon 9/11* (Washington, DC: U.S. Government Printing Office, 2007), 53–54, hereafter Goldberg, *Pentagon 9/11*.
4. Col Bradley H. Shumaker intvw, 2007 (Marine Corps History Division [MCHD], Quantico, VA).
5. George W. Bush, "Address to a Joint Session of Congress and the American People," 20 September 2001, White House Office of the Press Secretary Release, http://georgewbush-whitehouse.archives.gov/news/releases/2001/09/20010920-8.html, hereafter Bush, "Address to a Joint Session."
6. Goldberg, *Pentagon 9/11*, 10.
7. Mark D. St. Amand intvw, 18 December 2006 (MCHD, Quantico, VA).
8. Cdr William K. Saint Jr., USN, (Ret), e-mail to author, 27 September 2008 (MCHD, Quantico, VA).
9. Alan Levin, Marilyn Adams, and Blake Morrison, "Amid Terror, a Drastic Decision: Clear the Skies," *USA Today*, 12 August 2002.
10. LtCol Robert A. Ballard intvw with Fred H. Allison, 25 August 2004 (Oral HistColl, Gray Research Center, Quantico, VA).
11. Goldberg, *Pentagon 9/11*, 129.
12. Col Nathan S. Lowrey, "Marine Expeditionary Operations in Afghanistan: Task Force 58 during Operation Enduring Freedom September 2001–March 2002," (MS, MCHD, Quantico, VA, 2007), 19, hereafter Lowrey, "Marine Expeditionary Operations in Afghanistan."
13. Bush, "Remarks at Barksdale."
14. Senate Joint Resolution 23, 18 September 2001.
15. Lowrey, "Marine Expeditionary Operations in Afghanistan," 21.
16. Col Nicholas E. Reynolds (Ret), *Basrah, Baghdad, and Beyond: The U.S. Marine Corps in the Second Iraq War* (Annapolis, MD: Naval Institute Press, 2005), 2, hereafter Reynolds, *Basrah, Baghdad, and Beyond* (Naval Institute).
17. Lowrey, "Marine Expeditionary Operations in Afghanistan," 25.
18. Ibid., 27–28.
19. Bush, "Address to a Joint Session."
20. Lowrey, "Marine Expeditionary Operations in Afghanistan," 53.
21. LtCol Raymond C. Damm intvw with Fred H. Allison, 17 March 2002, (MCHD, Quantico, VA).
22. "Operation Enduring Freedom—Afghanistan," Global Security.org, 7 March 2005, http://www.globalsecurity.org/military/ops/enduring-freedom.htm.

Chapter 1

1. United States Marine Corps History Division, "Campaign Chronologies of the United States Marine Corps: Global War on Terrorism, 2001–2006," (MCHD, Quantico, VA).
2. 3d MAW Command Chronology (ComdC), January–June 2003 (Gray Research Center, Quantico, VA), hereafter 3d MAW ComdC.
3. Ibid.
4. Joseph J. Collins, *Choosing War: The Decision to Invade Iraq and Its Aftermath*, Institute for National Strategic Studies Occasional Paper 5 (Washington, DC: National Defense University Press, 2008), hereafter Collins, *Choosing War*.
5. Ibid.
6. United States Central Command, "Area of Responsibility," http://www.centcom.mil/.
7. United Nations Security Council, "Resolution 1441" (8 November 2002), http://www.un.org/depts/unmovic/new/documents/resolutions/s-res-1441.pdf.
8. William J. Clinton, "Statement on Signing the Iraq Liberation Act of 1998," 31 October 1998. Online by Gerhard Peters and John T. Woolley, *American Presidency Project*, http://www.presidency.ucsb.edu/ws/?pid=55205.
9. Reynolds, *Basrah, Baghdad, and Beyond* (Naval Institute), 1–14.
10. Baker Spring, "Operation Iraqi Freedom: Military Objectives Met," *Heritage Foundation*, 18 April 2003.
11. Department of Defense, "Goldwater-Nichols Reorganization Act of 1986," National Defense University Library.
12. Reynolds, *Basrah, Baghdad and Beyond* (Naval Institute), 7.
13. Ralph W. Donnelly, Gabrielle M. Neufeld, and Caro-

lyn A. Tyson, *A Chronology of the United States Marine Corps, 1947–1964*, vol. 3 (Washington, DC: Historical Division, Headquarters, U.S. Marine Corps [HQMC], 1971), 27.
14. Lowrey, "Marine Expeditionary Operations in Afghanistan," 65.
15. Reynolds, *Basrah, Baghdad, and Beyond* (Naval Institute), 3–4.
16. Department of the Navy, *Aviation Operations, MCWP 3-2* (Washington, DC: HQMC, 2000), chapter 2.
17. Col Nicholas E. Reynolds (Ret), "Command List," app. B in *U.S. Marines in Iraq, 2003: Basrah, Baghdad and Beyond*, (Washington, DC: MCHD, 2007), hereafter Reynolds, *Basrah, Baghdad, and Beyond* (MCHD).
18. Department of the Navy, *Marine Corps Operations, MDCP 1-0* (Washington, DC: HQMC, 2001), 3-10–3-15.
19. *Statement by Major General Stalder, United States Marine Corps, Deputy Commanding General I MEF, before the House Armed Services Committee, Subcommittee on Terrorism, Unconventional Threats and Capabilities on October 21, 2003, Concerning I MEF C4I during Operation Iraqi Freedom* (statement of MajGen Keith J. Stalder, USMC), hereafter Stalder statement.
20. LtGen James F. Amos intvw, 20 June 2007 (MCHD, Quantico, VA), hereafter Amos intvw, 20 June 2007.
21. Col Wade C. Hill, "A Drift to Joint: A Joint Warfare Question," 21 February 2004 (Quantico, VA: Marine Corps University).
22. MajGen James F. Amos intvw, 16 May 2003 (MCHD, Quantico, VA), hereafter Amos intvw, 16 May 2003.

Chapter 2

1. Lowrey, "Marine Expeditionary Operations in Afghanistan," 51.
2. Michael S. Groen, *With the 1st Marine Division in Iraq, 2003: No Greater Friend, No Worse Enemy*, occasional paper (Quantico, VA: MCHD, 2006), 1, hereafter Groen, *With the 1st Marine Division*.
3. Ibid.
4. Paul W. Westermeyer, *U.S. Marines in Battle: Al-Khafji, 28 January–1 February 1991* (Washington, DC: MCHD, 2008), 1, hereafter Westermeyer, *Al-Khafji*.
5. *The American Heritage Dictionary of the English Language*, 4th ed., (New York: Houghton Mifflin Company, 2004).
6. Reynolds, *Basrah, Baghdad, and Beyond* (Naval Institute), 17.
7. LtGen Wallace C. Gregson, "I Marine Expeditionary Force Summary of Action," in *U.S. Marines in Iraq, 2003: Anthology and Annotated Bibliography*, ed. and comp. Maj Christopher M. Kennedy, et al. (Washington, DC: MCHD, 2006), 4, hereafter Gregson, "I Marine Expeditionary Force"; hereafter Kennedy et al., *Iraq 2003 Anthology*.
8. U.S. Department of Defense, *MNF-I Fact Sheet OIF*, 20 March 2006.
9. U.S Department of State, *Bureau of Near Eastern Affairs*, June 2007.
10. Groen, *With the 1st Marine Division*, 5.
11. Ibid., 9.
12. Ibid., 20.
13. Ibid., 9.
14. Ibid., 96–97.
15. Department of the Navy, *Warfighting, MCDP 1* (Washington, DC: Headquarters U.S. Marine Corps, 1997), 72–76, hereafter DON, *Warfighting*.
16. LtCol Lawrence R. Roberts and Maj John P. Farnam, "Airborne Recon Supported Marines' Advances in Iraq," in Kennedy et al., *Iraq 2003 Anthology*, 145–47.
17. Groen, *With the 1st Marine Division*, 133.
18. DON, *Warfighting*, 93–95.

Chapter 3

1. Reynolds, *Basrah, Baghdad, and Beyond* (MCHD), 24–25.
2. Westermeyer, *Al-Khafji*, 1–3.
3. Harlan K. Ullman and James P. Wade, with L. A. "Bud" Edney, Fred M. Franks, Charles A. Horner, Jonathan T. Howe, and Keith Brendley, *Shock and Awe* (Washington, DC: National Defense University Press, 1996), hereafter Ullman and Wade, *Shock and Awe*.
4. Collins, *Choosing War*.
5. LtGen T. Michael Moseley, USAF, "Operation Iraqi Freedom—By the Numbers," U.S. Air Forces Central Command, Assessment and Analysis Division (Montgomery, AL: Maxwell Air Force Base, 2003), hereafter Moseley, "Operation Iraqi Freedom."
6. Operation Southern Focus, 1 March 2009, http://www.globalsecurity .org/military/ops/southern_focus.htm.
7. Andrew F. Krepinevich, "Operation Iraqi Freedom: A First-Blush Assessment" (Washington, DC: Center for Strategic and Budgetary Assessments, 2003).
8. Thomas R. Mockaitis, *The Iraq War: Learning from the Past, Adapting to the Present, and Planning for the Future* (Carlisle, PA: Strategic Studies Institute, 2007).
9. Reynolds, *Basrah, Baghdad, and Beyond* (MCHD), 43.
10. Ullman and Wade, *Shock and Awe*, xxiv.
11. Gregg Easterbrook, "JDAM the Torpedoes: The Weapons and Tactics We Will Use in Iraq," *Slate*, 5 March 2003, http://www.slate.com/id/2079667.
12. Maj J. R. McKay, "Mythology and the Air Campaign in the Liberation of Iraq," *Journal of Military and Strategic Studies* 7, no. 3 (2005): 1–7.
13. 3d MAW ComdC.
14. Department of the Navy, *Expeditionary Operations, MCDP 3* (Washington, DC: HQMC, 1998), 69–74, hereafter DON, *Expeditionary Operations*.
15. 3d MAW ComdC.
16. DON, *Expeditionary Operations*, 66–122.

17. Headquarters Marine Corps, Marine aviation website, https://hqinet001.hqmc.usmc.mil/AVN.
18. Amos intvw, 16 May 2003.
19. Richard A. Menton, "Airpower on the Maneuver Battlefield," *Marine Corps Gazette*, August 1989.
20. U.S. Air Force, *Counterland Operations, Air Force Doctrine Document 2-1.3* (Montgomery, AL: Maxwell Air Force Base, 2006), 5–7, 18.
21. Col L. Ross Roberts, *Ground Truth: The Implication of Joint Interdependence of Air and Ground Operations*, occasional paper no. 52, Center for Strategy and Technology (Montgomery, AL: Air University, 2006).
22. Maj David P. Westridge, "FMFM 1 and Its Implications for USMC Air," Command and Staff College (Quantico, VA: Marine Corps University, 1990).
23. Enduring Freedom Combat Assessment Team, "Aviation Ground Support," initial observation report (Quantico, VA, October 2003).
24. Amos intvw, 20 June 2007.
25. MACG-38 ComdC, January–June 2003 (Quantico, VA: Gray Research Center), hereafter MACG-38 ComdC.
26. Ibid.
27. 1st MarDiv ComdC, January–June 2003 (Quantico, VA: Gray Research Center), hereafter 1st MarDiv ComdC.
28. MAG-16 ComdC, January–June 2003 (Quantico, VA: Gray Research Center), hereafter MAG-16 ComdC.
29. 3d MAW ComdC.
30. I MEF ComdC, January–June 2003 (Quantico, VA: Gray Research Center), hereafter I MEF ComdC.
31. Ibid.
32. Joint Chiefs of Staff, *Joint Tactics, Techniques, and Procedures for Close Air Support (CAS), Joint Pub 3-09.3* (1995).
33. 4th MAW ComdC, January–June 2003 (Quantico, VA: Gray Research Center).
34. LtGen James F. Amos intvw, 6 June 2007 (Quantico, VA: MCHD).
35. I MEF ComdC.
36. LtGen James F. Amos intvw, 27 June 2006 (Quantico, VA: MCHD), hereafter Amos intvw, 27 June 2006.
37. Marine Corps Order P3500.37A, August 2005.
38. 3d MAW ComdC.
39. Amos intvw, 20 June 2007.
40. BGen Terry R. Robling intvw, 30 November 2006 (Quantico, VA: MCHD), hereafter Robling intvw, 30 November 2006.
41. Moseley, "Operation Iraqi Freedom."
42. Alexander M. Wathen, "The Miracle of Operational Iraqi Freedom Airspace Management," *Air & Space Power Journal–Chronicles Online Journal*, 4 October 2005.
43. MACG 38 ComdC.
44. http://www.globalsecurity.org.
45. 3d MAW ComdC.
46. Col Glibert B. Diaz intvw, 23 April 2003 (Quantico, VA: MCHD), hereafter Diaz intvw.
47. 3d MAW ComdC.
48. Diaz intvw.
49. CWO-2 Kraig A. Meyer intvw, 4 April 2003 (Quantico, VA: MCHD).

Chapter 4

1. LtCol Jeffrey P. Davis intvw, 27 January 2009 (Quantico, VA: MCHD).
2. MACG-38 ComdC.
3. Department of the Navy, *Direct Air Support Center Handbook, MCWP 3-25.5* (Washington, DC: HQMC, 2001).
4. MWCS-38 ComdC, January–June 2003 (Quantico, VA: Gray Research Center).
5. MACG-38 ComdC.
6. Ibid.
7. Ibid.
8. Ibid.
9. Col Jeffrey A. White intvw, 23 March 2003 (Quantico, VA: MCHD), hereafter White intvw.
10. LtCol Michael J. Kibler, "A Different Air War: Marine Air Control in Operation Iraqi Freedom," in Kennedy, et al., *Iraq 2003 Anthology*, 135, hereafter Kibler, "A Different Air War."
11. Rebecca Grant, "Marine Air in the Mainstream," *Air Force Magazine*, June 2004, http://www.afa.org.
12. LtGen James F. Amos intvw, 20 June 2006 (Quantico, VA: MCHD).
13. MWSG-37 ComdC, January–June 2003 (Quantico, VA: Gray Research Center), hereafter MWSG-37 ComdC.
14. Ibid.

Chapter 5

1. Col Juan G. Ayala intvw, 13 March 2003 (Quantico, VA: MCHD), hereafter Ayala intvw.
2. Col Stuart L. Knoll intvw, 5 March 2003 (Quantico, VA: MCHD).
3. 3d MAW ComdC.
4. U.S. Navy, "U.S. Navy Order of Battle: Operation Iraqi Freedom," http://navysite.de/navy/iraqi-freedom.htm, hereafter Navy Order of Battle.
5. MAG-13 ComdC, January–June 2003 (Quantico, VA: Gray Research Center), hereafter MAG-13 ComdC.
6. 2d MEB ComdC, January–June 2003 (Quantico, VA: Gray Research Center).
7. Navy Order of Battle.
8. U.S. Navy's Military Sealift Command, "Prepositioning (PM3)," http://www.msc.navy.mil/pm3.
9. 3d MAW ComdC, 15–16.
10. Ibid., 5.
11. Ibid., 19.
12. Ibid., 5.

13. Ibid.
14. Ayala intvw.
15. Col Michael D. Visconage, USMCR, and Maj Carroll N. Harris, USMCR, *Third Marine Aircraft Wing: Operation Iraqi Freedom* (Quantico, VA: Marine Corps Association, 2004), hereafter Visconage and Harris, *Third Marine Aircraft Wing*.
16. BGen Terry G. Robling intvw, 13 March 2003 (Quantico, VA: MCHD).
17. VMGR-234, OEF Standard Operating Procedures, 23 February 2003 (Quantico, VA: MCHD).
18. LtCdr Cindy Rodriguez, USN, et al., "Harrier Carriers Perform in Iraqi Freedom," in Kennedy et al., *Iraq 2003 Anthology*, 149–52.
19. Capt Jennifer M. Dolan intvw, 13 May 2003 (Quantico, VA: MCHD).
20. MAG-13 ComdC.
21. MAG-11 ComdC, January–June 2003 (Quantico, VA: Gray Research Center), hereafter MAG-11 ComdC.
22. MAG-13 ComdC.
23. MAG-16 ComdC.
24. 3d MAW ComdC.
25. MAG-29 ComdC, January–June 2003 (Quantico, VA: Gray Research Center), MAG-29 ComdC.
26. MajGen James F. Amos intvw, 27 June 2003 (Quantico, VA: MCHD), hereafter Amos intvw, 27 June 2003.
27. MWSS-272 ComdC, January–June 2003 (Quantico, VA: Gray Research Center); GySgt Ulysses L. Jackson intvw, 26 March 2003 (Quantico, VA: Gray Research Center).
28. Amos intvw, 27 June 2003.
29. Col Robert W. Hillery intvw, 23 April 2005 (Quantico, VA: MCHD).

Chapter 6

1. Reynolds, *Basrah, Baghdad and Beyond* (MCHD), 44–45.
2. Groen, *With the 1st Marine Division*, 114.
3. I MEF ComdC.
4. Groen, *With the 1st Marine Division*, 127.
5. Reynolds, *Basrah, Baghdad and Beyond* (MCHD), 55.
6. 3d MAW ComdC.
7. MACG-38 ComdC.
8. 3d MAW ComdC.
9. Stalder statement.
10. Amos intvw, 20 June 2007.
11. LtCol Wade C. Hall, "A Drift to Joint: A Joint Electronic Warfare Question," case study (Quantico, VA: Marine Corps Combat Development Command, 2004).
12. Sgt David A. Bryant, "Harriers Win the Day for Marine Corps Aviation during Operation Iraqi Freedom," *Marine Corps News*, 7 July 2003.
13. VMFA-533 ComdC, January–June 2003 (Quantico, VA: Gray Research Center), 8.
14. VMU-1 ComdC, January–June 2003 (Quantico, VA: Gray Research Center); VMU-2 ComdC, January–June 2003 (Quantico, VA: Gray Research Center), hereafter VMU-2 ComdC.
15. Groen, *With the 1st Marine Division*, 73.
16. Ibid.; Reynolds, *Basrah, Baghdad, and Beyond* (Naval Institute), 95.
17. 3d MAW ComdC.
18. Ibid.
19. I MEF ComdC.
20. 3d MAW ComdC.
21. Col Randolph D. Alles intvw, 23 March 2003 (Quantico, VA: MCHD).
22. 3d MAW ComdC.
23. Reynolds, *Basrah, Baghdad, and Beyond* (MCHD), 47.
24. Ibid., 55.

Chapter 7

1. George W. Bush, "President Says Saddam Hussein Must Leave Iraq within 48 Hours," 17 March 2003, White House Office of the Press Secretary Release, http://georgewbush-whitehouse.archives.gov/news/releases/2003/03/20030317-7.html.
2. Ibid.; SSgt Cindy Fisher, "Operation Iraqi Freedom—Marines Head into Iraq, Engage in Battle in March," *Marine Corps News*, 23 May 2003.
3. 1st MarDiv ComdC.
4. 3d MAW ComdC.
5. Visconage and Harris, *Third Marine Aircraft Wing*, 199.
6. Groen, *With the 1st Marine Division*, 142.
7. I MEF ComdC.
8. Ibid.
9. 3d MAW ComdC, 16.
10. Groen, *With the 1st Marine Division*, 54–56.
11. VMAQ-1 ComdC, January–June 2003 (Quantico, VA: Gray Research Center), hereafter VMAQ-1 ComdC.
12. I MEF ComdC.
13. Jay A. Stout, *Hammer from Above: Marine Air Combat Over Iraq*, (New York, NY: Presidio Press, 2005), 63–66, hereafter Stout, *Hammer from Above*.
14. White intvw.
15. Ibid.
16. MAG-29 ComdC.
17. Ibid.
18. Lindsay Murdoch, "Slaughter at Safwan Hill," *Sydney Morning Herald*, 22 March 2003.
19. VMFA(AW)-121 ComdC, January–June 2003 (Quantico, VA: Gray Research Center).
20. Stout, *Hammer from Above*.
21. Col Michael D. Visconage, Field History Diary, January–May 2003 (Quantico, VA: MCHD), hereafter Visconage, Field History Diary.
22. MAG-29 ComdC.

23. Richard S. Lowry, *Marines in the Garden of Eden: The Story of Seven Bloody Days In Iraq* (New York: Penguin, 2006), 91–93, hereafter Lowry, *Marines in the Garden.*
24. I MEF ComdC.
25. Capt Brian D. Laslie, USAF, review of *Cobra II: The Inside Story of the Invasion and Occupation of Iraq*, by Michael R. Gordon and Bernard E. Trainor *Air & Space Power Journal*, Spring 2007.
26. DON, *Warfighting*, 42–45.
27. I MEF ComdC.
28. Anthony H. Cordesman, *The Iraq War: A Working Chronology* (Washington, DC: Center for Strategic and International Studies, 2003), hereafter Cordesman, *Iraq War.*
29. Maj Scott A. Cooper intvw, 20 August 2008 (Quantico, VA: MCHD).
30. Groen, *With the 1st Marine Division*, 141.
31. 3d MAW ComdC.
32. Cordesman, *Iraq War.*
33. "Weapons of Mass Destruction (WMD), Tab A—Acronyms, Abbreviations, and Glossary," Global Security.org, http://www.globalsecurity.org/wmd/library/news/iraq/2000/scud_info/scud_info_taba.htm.
34. I MEF ComdC.
35. Ibid.
36. Cordesman, *Iraq War.*
37. Groen, *With the 1st Marine Division*, 142–43.
38. Gregson, "I Marine Expeditionary Force," 1.
39. Groen, *With the 1st Marine Division*, 142–43.
40. Gregson, "I Marine Expeditionary Force," 1.
41. VMFA(AW)-533 ComdC, January–June 2003 (Quantico, VA: Gray Research Center), hereafter VMFA(AW)-533 ComdC.
42. Ibid.; Stout, *Hammer from Above*, 62–76.
43. Groen, *With the 1st Marine Division*, 143.
44. Ibid.
45. Stout, *Hammer from Above*, 63–66.
46. VMFA(AW)-533 ComdC.
47. Groen, *With the 1st Marine Division*, 54.
48. Ibid., 9.
49. Col Ronnell R. McFarland intvw, 6 October 2004 (Quantico, VA: MCHD).
50. Kibler, "A Different Air War."

Chapter 8

1. MajGen Robin V. Brims intvw, 10 May 2003 (Quantico, VA: MCHD).
2. United States Marine Corps Operation Enduring Freedom Combat Assessment Team, "Operation Iraqi Freedom Report on Major Combat Operations," draft copy (Quantico, VA: Center for Lessons Learned).
3. Gregson, "I Marine Expeditionary Force," 1.
4. Groen, *With the 1st Marine Division*, 143.
5. Ibid., 154–59.
6. Bing West and MajGen Ray L. Smith (Ret), *The March Up: Taking Baghdad with the United States Marines* (New York: Random House, 2003), 15, hereafter West and Smith, *The March Up.*
7. HMLA-267 ComdC, January–June 2003 (Quantico, VA: Gray Research Center), hereafter HMLA-267 ComdC.
8. LtCol Stephen K. Heywood intvw, 6 April 2003 (Quantico, VA: MCHD).
9. MALS-39 ComdC, January–June 2003 (Quantico, VA: Gray Research Center).
10. Ibid.
11. VMA-311 ComdC, January–June 2003 (Quantico, VA: Gray Research Center), hereafter VMA-311 ComdC.
12. RCT-5 ComdC, January–June 2003 (Quantico, VA: Gray Research Center).
13. Gregson, "I Marine Expeditionary Force," 1.
14. 2d MEB, Task Force Tarawa ComdC, January–June 2003 (Quantico, VA: Gray Research Center), hereafter Task Force Tarawa ComdC.
15. 3d MAW ComdC.
16. Commander and Staff, MWSG-37, "No Farp Too Far!," in Kennedy et al., *Iraq 2003 Anthology*, 185–86, hereafter Commander and Staff, "No Farp."
17. LtCol Roger R. Machut intvw, 8 April 2003 (Quantico, VA: MCHD).
18. Cpl Jeff Hawk, "Profile: U.S. Marine Corps Chief Warrant Officer-4 Tom Cierley—Bulk Fueler Brings Bulk of Experience to Iraq," *Defend America News*, 5 January 2005.
19. VMU-2 ComdC.
20. Groen, *With the 1st Marine Division*, 219–20.
21. LtCol James R. Braden intvw, 23 April 2003 (Quantico, VA: MCHD), hereafter Braden intvw.
22. Stout, *Hammer from Above*, 72.
23. Braden intvw.
24. Stout, *Hammer from Above*, 72–75.
25. 3d MAW ComdC.
26. Ibid.
27. 15th MEU ComdC, January–June 2003 (Quantico, VA: Gray Research Center).
28. Maj James B. Hanlon intvw, 23 April 2005 (Quantico, VA: MCHD).
29. MACG-38 ComdC.
30. Ibid.
31. Kibler, "A Different Air War," 134.
32. 3d MAW Readiness Rates on 21Mar03, PowerPoint presentation.
33. 3d MAW ComdC, 16.
34. Gregson, "I Marine Expeditionary Force," 2.
35. Groen, *With the 1st Marine Division*, 56.
36. "RCT-5 Narrative Summary," 29 April 2003.
37. Cordesman, *Iraq War.*
38. Reynolds, *Basrah, Baghdad, and Beyond* (MCHD), 65.

39. Groen, *With the 1st Marine Division*, 174–78.
40. Amos intvw, 20 June 2007.

Chapter 9

1. Groen, *With the 1st Marine Division*, 180–82.
2. Col John R. Andrew Jr., USMCR, *U.S. Marines in Battle: An-Nasiriyah, 23 March–2 April 2003* (Washington, DC: MCHD), 14–15.
3. MajGen Richard F. Natonski intvw, 25 April 2003 (Quantico, VA: MCHD), hereafter Natonski intvw, 25 April 2003.
4. Groen, *With the 1st Marine Division*, 9.
5. MajGen Richard F. Natonski intvw, 20 December 2007 (Quantico, VA: MCHD).
6. Natonski intvw, 25 April 2003.
7. Reynolds, *Basrah, Baghdad, and Beyond* (MCHD), 70–75.
8. Stout, *Hammer from Above*, 108–10.
9. An-Nasiriyah OIF-I Close Air Support Panel intvw, 13 October 2006 (Quantico, VA: MCHD), hereafter Close Air Support Panel intvw.
10. Ibid.
11. Robert L. Sherrod, *History of Marine Corps Aviation in World War II* (Washington, DC: Combat Forces Press, 1952), 26–29.
12. Gen Robert E. Cushman Jr. intvw, 17 May 1974 (Quantico, VA: MCHD).
13. Maj Mike Huber intvw, 30 November 2006 (Quantico, VA: MCHD), hereafter Huber intvw.
14. Maj Mark D. Tobin intvw, 19 December 2006 (Quantico, VA: MCHD), hereafter Tobin intvw.
15. "U.S. CentCom Investigation of Suspected Friendly Fire Incident Near An-Nasiriyah, Iraq, 23 March 2003," (Tampa, FL: MacDill AFB, 6 March 2004), hereafter CentCom investigation.
16. Department of Defense, Press Release Number. 04-03-51, 29 March 2004, hereafter DOD press release.
17. I MEF Sitrep 221800Z March 2003 to 2231759Z March 2003.
18. DOD press release; Lowry, *Marines in the Garden*, 198, 211–13.
19. CentCom investigation.
20. Close Air Support Panel intvw.
21. Huber intvw.
22. Col Ronald L. Bailey intvw, 4 May 2006, (Quantico, VA: MCHD).
23. Marine Forces Europe, Landstuhl Regional Medical Center, Hospital Liaison Team PowerPoint brief, dated 2007.
24. MACG-38 ComdC.
25. Tobin intvw.
26. MACG-38 ComdC.
27. Stout, *Hammer from Above*, 108.
28. Gregson, "I Marine Expeditionary Force," 1–5.
29. Col Randolph D. Alles intvw, 20 April 2003 (Quantico, VA: MCHD).
30. Task Force Tarawa ComdC.
31. I MEF ComdC.
32. Task Force Tarawa ComdC.
33. Stout, *Hammer from Above*, 140; Robling intvw, 30 November 2006.
34. Visconage, Field History Diary.
35. 1st Marine Division Situation Report, 24 March 2003.
36. West and Smith, *The March Up*, 48.
37. BGen Terry G. Robling intvw, 23 April 2003 (Quantico, VA: MCHD).

Chapter 10

1. Groen, *With the 1st Marine Division*, 196–97; Stout, *Hammer from Above*, 171.
2. Reynolds, *Basrah, Baghdad, and Beyond* (MCHD), 77; Groen, *With the 1st Marine Division*, 199.
3. LtCol Jeffrey Acosta, "OIF Field History Journal," entry for 26–27 March 2003 (Quantico, VA: MCHD, 2003); LtCol Jerry G. Jamison intvw, 28 March 2003 (Quantico, VA: MCHD); LtCol Mark C. Graham intvw, 27 March 2003 (Quantico, VA: MCHD); SSgt Paul J. Morgado intvw, 24 April 2003 (Quantico, VA: MCHD); and Cdr Edward W. Hessell, USN, intvw, 27 March 2003 (Quantico, VA: MCHD).
4. Reynolds, *Basrah, Baghdad, and Beyond* (Naval Institute), 88.
5. Ibid., 89.
6. I MEF Sitrep 241800Z March 2003 to 251759Z March 2003.
7. Reynolds, *Basrah, Baghdad, and Beyond* (Naval Institute), 89.
8. Ibid., 90; BGen Edward G. Usher III intvw, 19 March 2003 (Quantico, VA: MCHD).
9. Ibid., 89–90.
10. Groen, *With the 1st Marine Division*, 215–17.
11. Ibid., 235
12. 3d MAW ComdC.
13. Reynolds, *Basrah, Baghdad, and Beyond* (Naval Institute), 90.
14. Groen, *With the 1st Marine Division*, chapter 5.
15. Ibid., 220–22.
16. 3d MAW ComdC.
17. MWSS-371 ComdC, January–June 2003 (Quantico, VA: Gray Research Center), hereafter MWSS-371 ComdC.
18. Cpl Jason Wachter intvw, 21 December 2003 (Quantico, VA: MCHD).
19. MACG-38 ComdC.
20. MWSS-371 ComdC.
21. Commander and Staff, "No Farp."
22. Col Charles G. Ikins, "Support to Operations Enduring and Iraqi Freedom: The Marine Corps Intelligence Activity," February 2002–September 2003, Headquarters, U.S. Marine Corps, 27.
23. Kibler, "A Different Air War," 138.

24. VMA-311 ComdC.
25. Sgt David A. Bryant, "Harriers Win the Day for Marine Corps Aviation during Operation Iraqi Freedom," *Marine Corps News*, 17 July 2003; VMA-211 ComdC, January–June 2003, (Quantico, VA: Gray Research Center), hereafter VMA-211 ComdC.
26. Reynolds, *Basrah, Baghdad, and Beyond* (Naval Institute), 89, 94.
27. Amos intvw, 20 June 2007.
28. Ibid.
29. Ibid.
30. Col John A. Toolan intvw, 10 December 2003 (Quantico, VA: MCHD).
31. Reynolds, *Basrah, Baghdad, and Beyond* (Naval Institute), 93.
32. I MEF ComdC.
33. Reynolds, *Basrah, Baghdad, and Beyond* (Naval Institute), 100–101.
34. MAG-16 ComdC, 7.
35. Task Force Tarawa ComdC, 6.
36. VMA-211 ComdC, 7.
37. Stout, *Hammer from Above*, 234–47.
38. HMH-465 ComdC, January–June 2003 (Quantico, VA: Gray Research Center), 4–5, 8.
39. VMA-211 ComdC, 7.
40. Ibid., 9.
41. Task Force Tarawa ComdC.
42. Reynolds, *Basrah, Baghdad, and Beyond* (MCHD), 89–94.
43. Ibid.
44. Reynolds, *Basrah, Baghdad, and Beyond* (Naval Institute), 92–93.
45. 3d MAW ComdC.
46. Moseley, "Operation Iraqi Freedom."
47. Groen, *With the 1st Marine Division*, 289.
48. Ibid., 297.
49. 3d MAW ComdC, 46–51.
50. Reynolds, *Basrah, Baghdad, and Beyond* (MCHD), 98.
51. Salman Pack, "Weapons of Mass Destruction," Global Security.org, http://www.globalsecurity.org.
52. BGen Robert F. Milstead Jr. notes to author, November 2008.
53. Amos intvw, 20 June 2007.
54. Ibid.
55. VMU-2 ComdC.
56. 3d MAW ComdC.
57. MWSG-37 ComdC.
58. Visconage, Field History Diary.
59. 3d MAW ComdC.
60. Amos intvw, 27 June 2006.
61. I MEF ComdC.
62. VMAQ-1 ComdC; VMAQ-2 ComdC, January–June 2003 (Quantico, VA: Gray Research Center).
63. LtCol Wade C. Hall, "A Drift to Joint," Command and Staff College, Marine Corps University, Quantico, VA, 23 February 2004.
64. Amos intvw, 27 June 2006.
65. 3d MAW ComdC.
66. Braden intvw.
67. Task Force Tarawa ComdC.
68. Groen, *With the 1st Marine Division*, 104–6.
69. Reynolds, *Basrah, Baghdad, and Beyond* (MCHD), 99–100.
70. Groen, *With the 1st Marine Division*, 305.
71. Reynolds, *Basrah, Baghdad, and Beyond* (MCHD), 101–3.
72. 3d MAW ComdC, 51.

Chapter 11

1. Reynolds, *Basrah, Baghdad, and Beyond* (MCHD), 103–5.
2. Gen Charles C. Krulak, "The Strategic Corporal: Leadership in the Three Block War," *Marines Corps Gazette*, January 1999.
3. Reynolds, *Basrah, Baghdad, and Beyond* (MCHD), 104–5.
4. I MEF Sequel (Post Hostility Operations) to I MEF OpOrd 1003V, 20 April 2003.
5. Reynolds, *Basrah, Baghdad, and Beyond* (MCHD), 104–5.
6. BGen John F. Kelly, "Tikrit, South to Babylon," in Kennedy et al., *Iraq 2003 Anthology*, 221–25.
7. I MEF ComdC.
8. VMA-211 ComdC.
9. Ibid.
10. Michael R. Gordon and LtGen Bernard E. Trainor (Ret), *Cobra II: The Inside Story of the Invasion and Occupation of Iraq* (New York: Pantheon, 2006), 438–56.
11. Reynolds, *Basrah, Baghdad, and Beyond* (MCHD), 107–11.
12. MACG-38 ComdC.
13. Reynolds, *Basrah, Baghdad, and Beyond* (MCHD), 109–11.
14. Groen, *With the 1st Marine Division*, 349–50; Maj Williams H. Holmes intvw, 24 April 2003 (Quantico, VA: MCHD); SSgt Paul J. Morgado intvw, 24 April 2003 (Quantico, VA: MCHD).
15. SSgt John C. DiDomenico, "VMGR-452 Delivers POWs to Freedom," *Marine Air Transporter*, July 2003, http://www.mcata.com/July%202003.pdf.
16. L. Paul Bremer, "What We Got Right in Iraq," *Washington Post*, 13 May 2007.
17. LtGen James T. Conway, "'Farther and Faster,'" in Kennedy et al., *Iraq 2003 Anthology*, 287.
18. LtGen Henry P. Osman intvw, 19 November 1993 (Quantico, VA: MCHD).

Chapter 12

1. Reynolds, *Basrah, Baghdad, and Beyond* (MCHD), 110–12.

2. 3d MAW ComdC.
3. Ibid.
4. Ibid.
5. I MEF, "EOS [expiration of obligated service]," July 2003, PowerPoint brief.
6. I MEF ComdC.
7. Maj Neil T. Allen, USAF, "Effects Based Assessment in the U.S. Air Force: Rhetoric or Reality?" (master's thesis, Air University, 2005).
8. Gregson, "I Marine Expeditionary Force," 5.
9. 3d MAW ComdC.
10. HMLA-267 ComdC.
11. 3dMAW ComdC.

Epilogue

1. Amos intvw, 16 May 2003.
2. MAG-11 ComdC.
3. Kibler, "A Different Air War," 137–38.
4. Reynolds, *Basrah, Baghdad, and Beyond* (MCHD), 55.

Appendix A
Command and Unit List

March–November 2003*

Marine Air-Ground Task Forces

U.S. Marine Forces Central Command [MarCentCom]
Commanding General: LtGen Earl B. Hailston

I Marine Expeditionary Force (Reinforced) [I MEF]
Commanding General: LtGen James T. Conway
Deputy: MajGen Keith J. Stalder

11th Marine Expeditionary Unit [11th MEU, TF Yankee]
Commanding Officer: Col Anthony M. Haslam

15th Marine Expeditionary Unit (Special Operations Capable) [15th MEU (SOC)]
Commanding Officer: Col Thomas D. Waldhauser

24th Marine Expeditionary Unit (Special Operations Capable) [24th MEU (SOC)]
Commanding Officers: Col Richard P. Mills
Col Ronald J. Johnson

Marine Aviation Combat Element

3d Marine Aircraft Wing [3d MAW]
Commanding General: MajGen James F. Amos
Assistant Wing Commander: BGen Terry G. Robling
Chief of Staff: Col Gerald A. Yingling Jr.

Marine Wing Headquarters Squadron 3 [MWHS-3]

Detachment, Marine Wing Headquarters Squadron 2 [Det, MWHS-2]

Atlantic Ordnance, Command Expeditionary Force [LantOrd, CmdExpedFor]

C-12 Detachment, Headquarters & Headquarters Squadron, MCAS Miramar, California [Det, HHS, NKX]

Marine Aircraft Group 11 (-) (Reinforced) [MAG-11]
Commanding Officer: Col Randolph D. Alles

*Basic sources are MarAdmin 507/03, various versions October–December 2003, with "Modifications to the I MEF Presidential Unit Citation," and command chronologies. Additions and/or corrections by Annette Amerman, a Reference Branch historian for the Marine Corps History Division, and Col Nicholas E. Reynolds's troop list of October 2004. Unit abbreviations are provided in brackets.

Marine Aviation Logistics Squadron 11 (-) (Reinforced) [MALS-11]

Marine Aviation Logistics Squadron 14 (-) [MALS-14]

Detachment, Marine Aviation Logistics Squadron 31 [Det, MALS-31]

Marine Aerial Refueler Transport Squadron 352 (-) (Reinforced) [VMGR-352]

Detachment, Marine Aerial Refueler Transport Squadron 234 [Det, VMGR-234]

Detachment, Marine Aerial Refueler Transport Squadron 452 [Det, VMGR-452]

Marine Fighter Attack Squadron 232 [VMFA-232]

Marine Fighter Attack Squadron 251 [VMFA-251]

Marine (All-Weather) Fighter Attack Squadron 121 [VMFA(AW)-121]

Marine (All-Weather) Fighter Attack Squadron 225 [VMFA(AW)-225]

Marine (All-Weather) Fighter Attack Squadron 533 [VMFA(AW)-533]

Marine Tactical Electronic Warfare Squadron 1 [VMAQ-1]

Marine Tactical Electronic Warfare Squadron 2 [VMAQ-2]

Marine Aircraft Group 13 (-) (Reinforced) [MAG-13]
Commanding Officer: Col Mark R. Savarese

Marine Aviation Logistics Squadron 13 (-) [MALS-13]

Marine Attack Squadron 211 (-) [VMA-211]

Marine Attack Squadron 214 [VMA-214]

Marine Attack Squadron 223 (-) [VMA-223]

Marine Attack Squadron 311 [VMA-311]

Marine Attack Squadron 542 [VMA-542]

Marine Aircraft Group 16 (-) (Reinforced) [MAG-16]
Commanding Officer: Col Stuart L. Knoll

Marine Aviation Logistics Squadron 16 (-) [MALS-16]

Detachment, Marine Aviation Logistics Squadron 26 [Det, MALS-26]

Marine Medium Helicopter Squadron 163 [HMM-163]

Marine Medium Helicopter Squadron 165 [HMM-165]

Marine Heavy Helicopter Squadron 462 [HMH-462]

Marine Heavy Helicopter Squadron 465 [HMH-465]

Marine Aircraft Group 29 (-) (Reinforced) [MAG-29]
Commanding Officer: Col Robert E. Milstead Jr.

Marine Aviation Logistics Squadron 29 (-) [MALS-29]

Marine Medium Helicopter Squadron 162 [HMM-162]

Marine Medium Helicopter Squadron 365 (-) [HMM-365]

Marine Heavy Helicopter Squadron 464 [HMH-464]

Marine Light Attack Helicopter Squadron 269 [HMLA-269]

Marine Aircraft Group 39 (-) (Reinforced) [MAG-39]
Commanding Officers: Col Richard W. Spencer
Col Kenneth P. Gardiner

Marine Aviation Logistics Squadron 39 (-) [MALS-39]

Marine Light Attack Helicopter Squadron 169 [HMLA-169]

Marine Light Attack Helicopter Squadron 267 [HMLA-267]

Marine Medium Helicopter Squadron 268 [HMM-268]

Marine Medium Helicopter Squadron 364 [HMM-364]

Marine Light Attack Helicopter Squadron 369 [HMLA-369]

Marine Air Control Group 38 (-) (Reinforced) [MACG-38]
Commanding Officer: Col Ronnell R. McFarland

Detachment, Marine Tactical Air Control Squadron 2 [Det, MTACS-2]

Marine Air Control Squadron 1 (Reinforced) [MACS-1]

Detachment, Marine Air Control Squadron 2 [Det, MACS-2]

Marine Wing Communications Squadron 28 (-) [MWCS-28]

Marine Wing Communications Squadron 38 (Reinforced) [MWCS-38]

Detachment, Marine Tactical Air Control Squadron 28 [Det, MTACS-28]

Marine Tactical Air Control Squadron 38 (Reinforced) [MTACS-38]

Detachment, Marine Tactical Air Control Squadron 48 [Det, MTACS-48]

Marine Air Support Squadron 1 [MASS-1]

Marine Air Support Squadron 3 (Reinforced) [MASS-3]

Battery B, 2d Low Altitude Air Defense Battalion [Btry B, 2d LAAD]

3d Low Altitude Air Defense Battalion [3d LAAD]

Detachment, Marine Air Support Squadron 6 [Det, MASS-6]

Marine Unmanned Aerial Vehicle Squadron 1 [VMU-1]

Marine Unmanned Aerial Vehicle Squadron 2 [VMU-2]

Marine Wing Support Group 37 (-) (Reinforced) [MWSG-37]
Commanding Officer: Col Michael C. Anderson

Marine Wing Support Squadron 271 [MWSS-271]

Marine Wing Support Squadron 272 [MWSS-272]

Marine Wing Support Squadron 371 [MWSS-371]

Marine Wing Support Squadron 372 [MWSS-372]

Marine Wing Support Squadron 373 [MWSS-373]

Company C, 1st Battalion, 24th Marines [Co C, 1st Bn, 24th Mar]

Military Police Detachment, 4th Marine Aircraft Wing [MP Det, 4th MAW]

Marine Ground Combat Element

1st Marine Division (Reinforced) [1st MarDiv]
Commanding General: MajGen James N. Mattis
Assistant Division Commander: BGen John F. Kelly

2d Marine Expeditionary Brigade [2d MEB, TF Tarawa]
Commanding General: BGen Richard F. Natonski

Marine Combat Service Support Element

Marine Logistics Command [MLC]
Commanding General: BGen Michael R. Lenhert

1st Force Service Support Group [1st FSSG]

Commanding Generals: BGen Edward G. Usher III
BGen Richard S. Kramlich

I Marine Expeditionary Force Engineer Group [I MEF EngGru]

Commanding Officer: RAdm Charles R. Kubic, USN

Marine Forces with Fifth and Sixth Fleets

26th Marine Expeditionary Unit (Special Operations Capable) [26th MEU (SOC)]

Commanding Officer: Col Andrew P. Frick

Marine Fighter Attack Squadron 115 [VMFA-115], USS *Harry S. Truman* (CVN 75)

Marine Fighter Attack Squadron 312 [VMFA-312], USS *Enterprise* (CVN 65)

Marine Fighter Attack Squadron 323 [VMFA-323], USS *Constellation* (CV 64)

Appendix B
Selected Glossary of Terms and Abbreviations

AAA–Antiaircraft artillery
AAV–Amphibious assault vehicle
ATARS–Advanced Tactical Air Reconnaissance System
ATF-East/West–Amphibious Task Force-East and -West
ATO–Air tasking order
AW–All-Weather
BCL–Battlefield coordination line
BCT–Brigade Combat Team
CAS–Close air support
CASEVAC–Casualty evacuation
CE–Command Element
CENTCOM–U.S. Central Command
CFACC–Coalition Forces Air Component Commander
CFLCC–Coalition Forces Land Component Commander
CG–Commanding General
CPA–Coalition Provisional Authority
CO–Commanding Officer
COC–Combat Operations Center
CONUS–Continental United States
CSS–Combat Service Support
CSSB–Combat Service Support Battalion
CSSC–Combat Service Support Company
CTF–Combined task force
DS–Direct support
EOD–Explosive ordnance disposal
FAC–Forward Air Controller
FAC(A)–Forward Air Controller (Airborne)
FARP–Forward arming and refueling point
FLOT–Forward linc of troops
FOB–Forward operating base
FPOL–Forward passage of lines
FRAGO–Fragmentary order
FSCL–Fire support coordination line
FSSG–Force Service Support Group
F/W–Fixed-wing
GCE–Ground Combat Element
HMH–Marine Heavy Helicopter Squadron
HMLA–Marine Light Attack Helicopter Squadron
HMM–Marine Medium Helicopter Squadron
HMMWV–High-Mobility, Multipurpose Wheeled Vehicle (Humvee)
JDAM–Joint Direct Attack Munitions
IFRC–International Federation of Red Cross and Red Crescent Societies

LAAD–Low Altitude Air Defense
LAR–Light Armored Reconnaissance
LASER–Light amplification by stimulated emission of radiation
MACG–Marine Air Control Group
MAG–Marine Aircraft Group
MAGTF–Marine Air-Ground Task Force
MALS–Marine Aviation Logistics Squadron
MASS–Marine Air Support Squadron
MAW–Marine Aircraft Wing
MAWTS–Marine Aviation Weapons and Tactics Squadron
MEB–Marine Expeditionary Brigade
MEF–Marine Expeditionary Force
MEU–Marine Expeditionary Unit
MMT– Marine Air Traffic Control Mobile Team
MOPP–Mission-oriented protective posture
MOS–Military occupational specialty
MPF–Maritime Prepositioning Force
MTACS–Marine Tactical Air Command Squadron
MWCS–Marine Wing Communications Squadron
MWSG–Marine Wing Support Group
MWSS–Marine Wing Support Squadron
NAI–Named area of interest
NORAD–North American Aerospace Defense Command
OEF–Operation Enduring Freedom
OIF–Operation Iraqi Freedom
OPLAN-Operational Plan
PAO–Public Affairs Officer
POW–Prisoner of war
RCT–Regimental Combat Team
ROC–Rehearsal of concept
ROE–Rules of engagement
ROZ–Restricted operation zone
RPG–Rocket-propelled grenade
R/W–Rotary-wing
SCAR–Strike Coordination and Reconnaissance
SeaBees–Navy construction unit (from CB for Construction Battalion)
SEAL–Sea-Air-Land team (U.S. Navy special operations force)
SOF–Special Operations Forces
SOP–Standard operating procedure
SSE–Sensitive site exploitation
SSM–Surface-to-surface missile
TAA–Tactical assembly area
TF–Task force
T/O–Table of organization
UAV–Unmanned aerial vehicle
VMA–Marine Attack Squadron
VMAQ–Marine Tactical Electronic Warfare Squadron
VMFA–Marine Fighter Attack Squadron
VMFA(AW)–Marine (All-Weather) Fighter Attack Squadron
VMGR–Marine Aerial Refueler Transport Squadron

VMU–Marine Unmanned Aerial Vehicle Squadron
WMD–Weapon of mass destruction

Appendix C
Chronology of Events

2003

6 January	3d Marine Aircraft Wing (3d MAW) Operational Plan (OPLAN) 1003V rehearsal of concept (ROC) drill at Marine Corps Air Station (MCAS) Miramar, California. Start of 3d MAW deployment in support of Operation Enduring Freedom/Operation Iraqi Freedom (OEF/OIF).
16 January	Maritime prepositioning force (MPF) offload begins.
17 January	Amphibious Task Force–West (ATF-W) sets sail from San Diego, California.
23 January	Wing advanced echelon deployed to Kuwait.
28 January	Select Marine Corps Reserve deploying units activated.
February	MPF offload complete (11 ships total).
10 February	Wing main body deployed to Kuwait.
13 February	Select Marine Corps Reserve CONUS generating forces activated.
20 February	G-5 published 3d MAW OPLAN 1003V.
21 February	Wing rear deployed to Kuwait.
28 February	Wing representatives attended I Marine Expeditionary Force (I MEF) ROC drill at Camp Commando, Kuwait.
5 March	Wing operational planning team stood up at al-Jaber Air Base in Kuwait and representatives attended force service support group ROC drill at Tactical Assembly Area Coyote.
19 March	OIF commences; OPLAN 1003V rules of engagement (ROE) in effect. 67 sorties flown. 427 untasked ground alert sorties. 28 Operation Southern Watch sorties flown. 10 fixed-wing (f/w) and 5 unmanned aerial vehicle (UAV) I MEF targeting collection sorties. 202 f/w and rotary-wing (r/w) close air support/forward air controller (airborne) (CAS/FAC[A]) I MEF ground alert sorties. 74 assault support I MEF ground alert missions. 12 aerial refueling ground alert missions.

	4 KC-130s staged forward at Forward Operating Base (FOB) Joe Foss. Air strikes begin on Safwan, Iraq; assault is aborted because of weather and rescheduled.
20 March	259 missions flown. 156 untasked ground alert sorties. 24 Coalition Forces Air Component Commander (CFACC) missions. 235 I MEF missions. 20 of 22 I MEF reconnaissance missions. Maintained constant airborne CAS coverage supporting Regimental Combat Team 5 (RCT-5) and multiple CAS missions in support of seizure of Division Objective 3. Maintained constant f/w FAC(A) coverage for both RCT-5 and RCT-7, flew f/w counter fire for RCT-7 during movement across line of departure. Several missions diverted by CFACC to strike CSSC-3 missiles near Basrah, and S-60 antiaircraft guns in the U.S. Army's V Corps zone. Began shaping effort against Iraqi second echelon forces. Safwan assault completed.
21 March	309 missions flown. 116 untasked ground alert sorties. 8 CFACC missions. 301 I MEF missions. 22 of 25 I MEF collections reconnaissance missions. Successfully prosecuted time-sensitive target, destroying residence with confirmation that Ali Hasan al-Majid was inside at the time of impact. Provided 8 f/w CAS sorties supporting United Kingdom (UK) forces engaged at al-Faw. Flew 3 casualty evacuation (CASEVAC), 12 general logistics support, 4 ammo resupply, 4 fuel resupply, and 2 external resupply missions. Completed 16 KC-130 3d MAW aerial refueling missions. Pushed Marine wing support squadron (MWSS) and Marine air traffic control mobile team (MMT) units forward to establish forward arming and refueling points (FARPs) at Safwan and Jalibah.
22 March	351 missions flown. 111 untasked ground alert sorties. 20 CFACC missions. 331 I MEF missions. 8 advanced tactical air reconnaissance system (ATARS) I MEF collections missions. Flew CAS in support of UK forces engaged near Umm Qasr. Provided coverage for 1st Marine Division (1st MarDiv) crossing of Euphrates River. Directed 4 UK Harriers and 6 USAF A-10s to support TF Tarawa at an-Nasiriyah. Executed 6 CASEVAC, 11 general logistics support, 2 ammo resupply, and 4 fuel resupply missions. Completed 15 KC-130 3d MAW aerial refueling missions.
23 March	317 missions flown. 133 untasked ground alert sorties. 9 CFACC missions. 308 I MEF missions. 10 ATARS I MEF collections missions.

Provided CAS for UK near Basrah and TF Tarawa in an-Nasiriyah.
CFACC provided multiple A-10s to support I MEF.
Inserted 3 force reconnaissance teams into named areas of interest (NAIs) 14, 16, and 18.
Executed CASEVAC, 11 general logistics support, 2 ammo resupply, and 4 fuel resupply missions.
Completed 13 KC-130 3d MAW aerial refueling missions.

24 March

301 missions flown.
141 untasked ground alert sorties.
13 CFACC missions.
288 I MEF missions.
10 ATARS I MEF collections missions.
Weather hampered operations in central and northern I MEF areas of operation.
Continued interdiction missions against Iraqi *6th* and *10th Divisions* near Basrah and al-Amarah.
Limited VMU missions, as units moved north in support of 1st MarDiv. Marine Unmanned Aerial Vehicle Squadron 1 (VMU-1) displaced to Camden Yards FARP. VMU-2 stopped at PacBell FARP.

25 March

Weather hampered flight operations in theater for 3 days.
3 missions flown.
114 untasked ground alert sorties.
13 CFACC missions.
177 I MEF missions.
10 ATARS I MEF collections missions.
Clearing began in mid-afternoon, with only limited flights before 1200Z.
Opened Wrigley (Highway 1) FARP.

26 March

151 missions flown.
119 untasked ground alert sorties.
13 CFACC missions.
36 I MEF missions.
2 ATARS I MEF collections missions.
Weather impeded sustained operations.
Coordinated Coalition aircraft from other bases.

27 March

194 missions flown.
108 untasked ground alert sorties.
10 CFACC missions.
184 I MEF missions.
4 ATARS I MEF collections missions.
VMU-1 reassigned to TF Tarawa.
3d MAW deployment in support of OIF completed.

28 March

371 missions flown.
73 untasked ground alert sorties.
11 CFACC missions.
350 I MEF missions.
10 ATARS I MEF collections missions.

Accepted operational control of Marine Medium Helicopter Squadron 263 (HMM-263) (Rein) from 24th Marine Expeditionary Unit (Special Operations Capable) (24th MEU [SOC]).

29 March	341 missions flown. 111 untasked ground alert sorties. 19 CFACC missions. 322 I MEF missions. 10 ATARS I MEF collections missions.
30 March	277 missions flown. 133 untasked ground alert sorties. 19 CFACC missions. 258 I MEF missions. 8 ATARS I MEF collections missions. VMU-1 and VMU-2 each flew 6 missions. Flew CAS in support of RCT-1 and TF Tarawa. Significant progress in buildup of FARPs throughout the I MEF areas of operation. MWSS-271 moved to Qualcom FARP.
31 March	308 missions flown. 122 untasked ground alert sorties. 20 CFACC missions. 288 I MEF missions. 10 ATARS I MEF collections missions. 77 common source 3d MAW sorties. Moved VMU-1 to Fenway FARP to support 1st MarDiv.
1 April	320 missions flown. 118 untasked ground alert sorties. 20 CFACC missions. 300 I MEF missions. 12 ATARS I MEF collections missions. 86 common source 3d MAW sorties.
2 April	336 missions flown. 126 untasked ground alert sorties. 17 CFACC missions. 319 I MEF missions. 14 ATARS I MEF collections missions. 42 common source 3d MAW sorties. Camden FARP closed. VMU-1 operating from Fenway FARP, and VMU-2 operating from QualCom FARP.
3 April	319 missions flown. 110 untasked ground alert sorties. 17 CFACC missions. 302 I MEF missions. 14 ATARS I MEF collections missions. 38 common source 3d MAW sorties.

Opened Three Rivers FOB near an-Numaniyah.
VMU-1 moved from Fenway to Three Rivers FOB.
Discontinued active engagement of enemy forces east of 045.07 Easting.

4 April
291 missions flown.
164 untasked ground alert sorties.
20 CFACC missions.
271 I MEF missions.
14 ATARS I MEF collections missions.
44 common source 3d MAW sorties.

5 April
262 missions flown.
163 untasked ground alert sorties.
77 CFACC missions.
185 I MEF missions.
14 ATARS I MEF collections missions.

6 April
279 missions flown.
174 untasked ground alert sorties.
89 CFACC missions.
190 I MEF missions.
14 ATARS I MEF collections missions.
VMU-1 and VMU-2 both operated at Three Rivers FOB and flew 6 and 4 missions, respectively.

7 April
244 missions flown.
156 untasked ground alert sorties.
89 CFACC missions.
155 I MEF missions.
14 ATARS I MEF collections missions.
VMU-1 operated at Fenway FARP; VMU-2 moved to Yankee FOB.

8 April
199 missions flown.
146 untasked ground alert sorties.
44 CFACC missions.
155 I MEF missions.
11 ATARS I MEF collections missions.
KC-130 operations at Yankee FOB.
VMU-1 operated at Qalat Sikar Air Base, flying 4 TF Tarawa missions; and VMU-2 operated at Yankee FOB, flying 4 1st MarDiv sorties.

9 April
183 missions flown.
154 untasked ground alert sorties.
64 CFACC missions.
119 I MEF missions.
13 ATARS I MEF collections missions.

10 April
198 missions flown.
160 untasked ground alert sorties.
44 CFACC missions.

	154 I MEF missions. 13 ATARS I MEF collections missions. VMU-1 flew out of Three Rivers FOB; VMU-2 and KC-130 conducted operations at Yankee FOB. Conducted 7 CASEVAC missions.
11 April	174 missions flown. 160 untasked ground alert sorties. 57 CFACC missions. 117 I MEF missions. 13 ATARS I MEF collections missions.
12 April	160 missions flown. 183 untasked ground alert sorties. 57 CFACC missions. 1 I MEF mission. 13 ATARS I MEF collections missions. VMU-1 flew 3 TF Tarawa missions and VMU-2 flew 4.
13 April	161 missions flown. 188 untasked ground alert sorties. 30 CFACC missions. 131 I MEF missions. 13 ATARS I MEF collections missions. Ebbets FARP closed.
14 April	180 missions flown. 178 untasked ground alert sorties. 43 CFACC missions. 131 I MEF missions. 12 ATARS I MEF collections missions.
15 April	166 missions flown. 197 untasked ground alert sorties. 41 CFACC missions. 125 I MEF missions. 14 ATARS I MEF collections missions.
1 May	End of major OIF combat.
5 May	3d MAW redeployment from OIF began.
24 May	Advanced echelon redeployed from OIF.
6 June	MajGen James A. Amos, 3d MAW, and staff returned to MCAS Miramar. BGen Terry G. Robling assumed duties as 3d MAW (forward) commanding general.

Appendix D
Unit Citation

THE SECRETARY OF THE NAVY
WASHINGTON, D.C. 20350-1000
3 November 2003

The President of the United States takes pleasure in presenting the PRESIDENTIAL UNIT CITATION to

I MARINE EXPEDITIONARY FORCE

for service as set forth in the following

CITATION:

For extraordinary heroism and outstanding performance in action against enemy forces in support of Operation IRAQI FREEDOM from 21 March to 24 April 2003. During this period, I Marine Expeditionary Force (MEF) (REIN) conducted the **longest sequence** of coordinated combined arms overland attacks in the history of the Marine Corps. From the border between Kuwait and Iraq, to the culmination of hostilities north of Baghdad, I MEF advanced nearly 800 kilometers under sustained and heavy combat. Utilizing the devastating combat power of organic aviation assets, coupled with the awesome power resident in the ground combat elements, and maintaining momentum through the herculean efforts of combat service support elements, I MEF destroyed nine Iraqi Divisions. This awesome display of combat power was accomplished while simultaneously freeing the Iraqi people from more than 30 years of oppression and reestablishing basic infrastructure in the country. During the 33 days of combat, to the transition to civil-military operations, I MEF sustained a tempo of operations never before seen on the modern battlefield, conducting four major river crossings, maintaining the initiative, and sustaining forces. The ferocity and duration of the campaign was made possible through the skills and determination of the Soldiers, Sailors, Airmen, Marines, and Coalition Partners comprising I MEF at all levels, all echelons, and in all occupational fields. By their outstanding courage, aggressive fighting spirit, and untiring devotion to duty, the officers and enlisted personnel of I Marine Expeditionary Force (REIN) reflected great credit upon themselves and upheld the highest traditions of the Marine Corps and the United States Naval Service.

For the President,

Secretary of the Navy

Appendix E
3d MAW Aircraft Deployed during Operation Iraqi Freedom I

McDonnell Douglas AV-8B Harrier II

This single-engine light attack aircraft is primarily known for its short takeoff and landing ability, which allows the fixed-wing high-speed plane to be forward based and close to the front-line warfighters. Its six wing stations can accommodate AIM-9 Sidewinder air-to-air missiles and an assortment of air-to-ground weapons, external fuel tanks, and AGM-65 Maverick missiles; one centerline fuselage station can carry an AN/ALQ-164 DECM (deception electronic countermeasures) pod. Armed with a Litening II targeting pod, the Harrier can deliver GBU-12 and GBU-16 laser-guided bombs with pinpoint accuracy. A GAU-12 25mm six-barrel gun pod and accompanying 300-round ammunition pod can also be mounted either side of centerline. The AV-8BII+ features APG-65 radar common to the F/A-18, as well as all systems and features of the AV-8B Harrier II.

Northrup Grumman EA-6B Prowler

The primary function of the Prowler is to engage in airborne electronic warfare support to Fleet Marine Forces. Operated by a crew of four, it features an ALQ-99 onboard system to collect tactical electronic order of battle data, which can be recorded and processed after missions to provide updates to various orders of battle. Its ALQ-99 tactical jamming system variation is used to provide active radar jamming support to assault support and attack aircraft, as well as ground units, while the aircraft's HARM (high-speed antiradiation missile) capability provides suppression of enemy air defenses. The USQ-113 communications jammer can detect and jam a wide range of communication frequencies to further degrade air defense and ground units' capabilities.

McDonnell Douglas F/A-18A and C Hornet

The single-seat A- and C-model Hornet is a twin-engine strike fighter designed to intercept and destroy enemy aircraft under all-weather conditions and attack and destroy surface targets. It features nine external weapon stations that can accommodate up to 13,700 pounds of weapons, sensors, and fuel. The C model, introduced in 1987, incorporated upgraded radar and avionics, as well as the capacity to carry new missiles, including the AIM-120 AMRAAM (advanced medium-range air-to-air missile) and AGM-65 Maverick and AGM-84 Harpoon air-to-surface missiles. Other improvements include the Martin-Baker NACES (Navy aircrew common ejection seat), a self-protection jammer, and synthetic aperture ground-mapping radar that enables the pilot to locate targets in poor visibility. Night attack abilities were improved in 1989 with the addition of the Hughes AN/AAR-50 thermal navigation pod, the Loral AN/AAS-38 NITE Hawk forward-looking infrared array targeting pod, night-vision goggles, and full-color multifunction displays and a color moving map.

McDonnell Douglas F/A-18D Hornet

Unlike the Navy, which uses this two-seat version of the F/A-18C as a trainer, the Marines employ the D model tactically as a strike fighter, but also as a forward air controller (airborne)/tactical air controller (airborne) and reconnaissance aircraft. In addition, its night attack suite allows pilots of the F/A-18D to conduct operations below weather at low altitude using night-vision goggles and forward-looking infrared radar systems. The D model, like the C, can carry up to 13,700 pounds of external ordnance, including a variety of sensor pods and precision-guided weapons on nine external hard points: two wingtip stations for AIM-9 Sidewinder air-to-air missiles and four underwing and three fuselage stations. Air-to-ground weapons include all GBU series bombs; JSOW, JDAM, and Mk80 series general-purpose bombs; and CBU-59 cluster bombs. An M61 Vulcan 6-barrel 20mm rotary cannon with 520 rounds of ammunition is internally mounted in the nose. Production of this Hornet was discontinued, and the last F/A-18D was delivered to the U.S. Marine Corps in 2000.

Lockheed KC-130F/R/T Hercules

The Hercules is a multirole, multimission tactical tanker/transport that provides the support required by Marine air-ground task forces. This versatile asset can land and take off on short, unimproved runways and provides in-flight refueling to tactical aircraft and helicopters, as well as rapid ground refueling when required. The Hercules also provides aerial troop and cargo delivery, emergency resupply into unimproved landing zones within the objective or battle area, airborne direct air support center, emergency medical evacuation, tactical insertion of combat troops and equipment, evacuation missions, and support as required of special operations capable forces.

IAI/AAI RQ-2A Pioneer

The Pioneer was the first tactical battlefield unmanned aerial vehicle (UAV) in service with the U.S. military. Israel Aircraft Industries began its development in 1984 with deliveries to the Marine Corps in July 1987. The UAV provides reconnaissance support to the Marine air-ground task force, which includes surveillance of designated areas of interest, reconnaissance of helicopter approach and retirement lanes, adjusting indirect fire support, support rear area security, and battle damage assessment. The Pioneer UAV is equipped with an inertial navigation system and a two-way C-band datalink. Although it can fly a preprogrammed mission, the craft is operated by the manual remote control using video provided by the datalink; maximum range for the line-of-sight datalink is 100 nautical miles. The RQ-2A can be equipped with a wide variety of mission payloads, primarily the Wescam DS-12 EO/IR (electro-optical/infrared) sensor.

Bell UH-1N Iroquois (nicknamed the "Huey" or "Twin Huey")

The Twin Huey is a twin-engine helicopter that can be outfitted to support such operations as command and control with a specialized communication package, arms coordination, assault, medical evacuation for up to six litter patients and one medical attendant, external cargo, search and rescue using an external hoist, and reconnaissance and reconnaissance support. The Huey provides utility combat helicopter support to the landing force commander during ship-to-shore movement and in subsequent operations ashore. Considered the most widely used helicopter in the world, the Huey is armed with either the M240 7.62mm machine gun, GAU-16/A .50-caliber machine gun, or the GAU-17/A 7.62mm minigun; all three weapons systems are crew-served. The helicopter can also carry two 7- or 19-shot 2.75-inch rocket pods.

Bell AH-1W Sea Cobra

The Sea Cobra (aka Super Cobra or "Whiskey Cobra") is a Marine Corps attack helicopter capable of operating day or night and with limited visibility. The two-person, tandem-seat, twin-engine aircraft is capable of land- or sea-based operations. It provides en route escort for assault helicopters; fire support and security for forward and rear area forces; point target/antiarmor; antihelicopter, armed escort; supporting arms control and coordination; point- and limited-area air defense from enemy fixed-wing aircraft; and armed and visual reconnaissance. The Sea Cobra is armed with a 20mm M197 three-barrel Gatling-type cannon with 750 rounds in a nose-mounted M97 turret, and four external stub-wing stations that can launch 2.75- or 5-inch rockets and precision-guided weapons, including BGM-71 TOW and AGM-114 Hellfire point target/antiarmor, AIM-9 Sidewinder antiair, and AGM-122 Sidearm antiradar missiles.

Boeing CH-46E Sea Knight

The Sea Knight is a medium-lift tandem rotor helicopter that provides all-weather, day or night assault transport of combat troops, supplies, and equipment. It also provides support for evacuation operations and other maritime special operations, over-water search-and-rescue augmentation, mobile FARPs, and aeromedical evacuation of casualties from the field to suitable medical facilities. The helicopter may be armed with up to two door-mounted GAU-15/A .50-caliber machine guns and a ramp-mounted M240D 7.62mm machine gun. The CH-46 was first procured in 1964, and the updated E model was produced in 1978. Although the craft continues to serve the Marine Corps in combat and peacetime environments, it is undergoing a phased replacement by Bell Boeing MV-22 Osprey tiltrotor aircraft.

Sikorsky CH-53D Sea Stallion

The Sea Stallion is a medium-lift helicopter designed to transport up to seven tons of equipment, supplies, and personnel. Capable of both internal and external supply transportation, the twin-engine D model is shipboard compatible and capable of operating in adverse weather conditions both day and night. Twin-turbine engines turn a single, six-bladed main rotor, which has an automatic blade folding system. An automatic flight control system reduces pilot fatigue on long missions. The CH-53 is capable of emergency water landing and takeoff. The Sea Stallion can carry 38 combat-equipped troops or 24 litter patients. It is armed with two GAU-15/A .50-caliber machine guns.

Sikorsky CH-53E Super Stallion

Despite its obvious relationship to previous versions of the CH-53, the E model is a totally different aircraft. The third engine, a seventh rotor blade, increased rotor diameter, and overall increased size bought the Marines—at 16-plus tons—more than double the 7-ton lift capacity of the D model. The Super Stallion normally seats 37 passengers, but it carries provisions for 55; it can carry external loads at increased airspeeds due to the stability achieved with the dual-point hook system. As the Marine Corps' heavy-lift helicopter, it is compatible with most amphibious ships and is carried routinely on board LHA (landing, helicopter, assault) and LHD (landing, helicopter, dock) ships. The CH-53E is equipped with a refueling probe and can be refueled in flight, giving the helicopter indefinite range. A dual digital automatic flight control system provides exceptional flying qualities in all flight modes and an engine anti-ice system that allows all-weather capability. Precision navigation is provided by an integrated global positioning system and further augmented by the night-vision system's forward-looking infrared sensor. Like the CH-53D, the Super Stallion carries two GAU-15/A .50-caliber machine guns.

Appendix F
3d MAW Statistics

Table 1. Summary of Flight Hours and Mishaps, 2003

Month	Monthly flight hours	Fiscal year flight hours	Aircraft mishap rate
January	8,033.7	23,627.1	9.0
February	8,841.9	32,469.0	7.2
March	16,589.3	49,058.3	6.0
April	17,803.2	66,861.5	5.1
May	10,200.4	77,061.9	4.5
June	9,466.0	86,527.9	3.9

Table 2. Summary of Class "A" Mishaps, 2003

Date	Squadron	Details
21 March	HMM-268	CH-46 crashed
25 March	VMU-1	Unmanned aerial vehicle crashed on landing
30 March	HMLA-169	UH-1N crashed
4 April	HMM-364	AH-1W crashed one-half-mile from landing zone
5 April	HMLA-267	AH-1W crashed, struck a tower while engaging enemy
14 April	HMLA-169	AH-1W destroyed by secondary explosions

Table 3. Daily Sorties Flown by 3d MAW, 2003

Date	Total	I MEF	Air Force CFACC
19 March	67	N/A	N/A
20 March	259	235	24
21 March	309	301	8
22 March	351	331	20
23 March	317	308	9
24 March	301	288	13
25 March	183	N/A	N/A
26 March	151	138	13
27 March	194	184	10
28 March	361	350	11
29 March	241	322	19
30 March	277	258	19
31 March	308	288	20
1 April	320	300	20
2 April	336	319	17
3 April	319	302	17
4 April	291	271	20
5 April	262	185	77
6 April	279	190	89
7 April	244	155	89
8 April	199	155	44
9 April	183	119	64
10 April	198	154	44
11 April	174	117	57
12 April	160	103	57
13 April	161	131	30
14 April	174	131	43
15 April	166	125	41

Note: Beginning on 5 April 2003, as the Coalition circle around Baghdad tightened, 3d MAW increased the number of sorties flown in support of the U.S. Air Force.

Table 4. Sortie Summary by Aircraft, 19-31 March 2003

	19	20	21	22	23	24	25	26	27	28	29	30	31	Total
AH-1W	2	23	22	36	23	22	15	0	6	25	16	11	16	217
AV-8B	16	56	121	118	92	88	32	43	38	101	95	76	75	951
CH-46E	2	30	7	19	23	25	16	3	13	41	31	35	27	272
CH-53E	2	19	23	20	28	16	8	2	4	26	26	16	23	213
EA-6B	12	12	13	14	9	13	6	10	10	11	11	12	12	145
F/A-18C	14	35	38	44	35	42	36	32	34	44	57	43	51	505
F/A-18D	12	49	46	70	68	69	44	32	54	78	68	54	58	702
KC-130	1	12	21	21	23	21	18	20	27	23	22	25	25	259
RQ-2B	5	10	12	1	0	0	0	1	4	5	5	9	9	61
UH-1N	1	13	6	8	14	5	8	3	4	7	10	4	6	89
Totals:	67	259	309	351	315	301	183	146	194	361	341	285	302	3,414

Source: 3d MAW PowerPoint briefing (2003), titled "Op Plan 1003V Sorties Flown 2003."

Note: Due to poor flying conditions and a sandstorm, 3d MAW flew fewer sorties on 19 and 25–26 March, which significantly limited rotary-wing air support on those days.

Back Cover: The device reproduced on the back cover is the oldest military insignia in continuous use in the United States. It first appeared, as shown here, on Marine Corps buttons adopted in 1804. With the stars changed to five points, the device has continued on Marine Corps buttons to the present day.

Made in the USA
Monee, IL
04 November 2026

67384745R00125